软件工程应用型专业“十三五”规划系列教材

天津工业大学软件学院
融创软通公司教育培训部　联合编写

Java EE 轻量级框架应用与开发

——Spring + Spring MVC + MyBatis

任淑霞 / 主　编
张建军　陈香凝 / 副主编

内容提要

本书主要包括3个部分，分别是Spring框架、Spring MVC框架和MyBatis框架。Spring+Spring MVC+MyBatis已经成了Java互联网时代的主流框架，而MySQL也已经成为主流的数据库，这三个框架已经构成了当前Java后端开发最流行、最核心的框架技术，也是当前就业者应会必会的技术，实用性很强。

本书基于企业的实际需求，通过将繁杂的框架知识点融入实际案例，把所有章节重点技术进行贯穿，且每章案例代码会层层迭代不断完善，最终形成一个完整的系统。同时各个章节都包含小结，“经典面试题”和“跟我上机”等内容，让读者技术更加扎实，增加找工作面试成功率，从而提高就业竞争力。

本书可作为高等学校计算机及软件相关专业Java课程高级阶段的教材，也可供从事基于Java的Web应用开发技术人员学习参考等。

图书在版编目（CIP）数据

Java EE轻量级框架应用与开发：Spring+Spring MVC+MyBatis/任淑霞主编.—天津：天津大学出版社，2019.1（2021.8重印）
校企协同软件工程应用型专业“十三五”实训规划系列教材
ISBN 978-7-5618-6339-8

Ⅰ.①J… Ⅱ.①任… Ⅲ.①JAVA语言－程序设计－高等学校－教材 Ⅳ.①TP312.8

中国版本图书馆CIP数据核字(2019)第012650号

Java EE Qingliangji Kuangjia Yingyong yu Kaifa——Spring+Spring MVC+MyBatis

出版发行　天津大学出版社
地　　址　天津市卫津路92号天津大学内（邮编：300072）
电　　话　发行部：022-27403647
网　　址　publish.tju.edu.cn
印　　刷　北京虎彩文化传播有限公司
经　　销　全国各地新华书店
开　　本　185mm×260mm
印　　张　19.75
字　　数　500千
版　　次　2019年1月第1版
印　　次　2021年8月第2次
定　　价　52.00元

前　言

本教材属于校企协同软件工程应用型专业实训系列丛书，是天津工业大学计算机科学与软件学院和融创软通公司的多位教师近12年校企协同育人过程的经验总结和不断提炼的教学成果。

本书编写背景

在Web应用的开发过程中，开发框架的选择非常重要。一个好的开发框架能够加速Web应用的开发速度，降低开发成本，减少开发人员的工作量，同时能够使Web应用具有良好的扩展性和移植性。基于Spring+Spring MVC+MyBatis(SSM)的框架凭借良好的性能和较快的开发效率，逐渐成为主流的Java EE Web应用开发框架组合。Spring MVC是一个基于MVC的框架，其主要负责表现层的功能，比如响应请求。Spring框架主要起到容器的功能，整合了Spring MVC和MyBatis，实现层与层之间的解耦，同时使业务逻辑更加清晰。MyBatis框架主要负责的是数据持久层，完成和数据库的相关操作。本书为开发基于Java EE的企业级大型应用提供了理论指导。

阅读本书所需的基础知识

SSM(Spring+Spring MVC+MyBatis)是目前较为主流的企业级架构方案，在软件开发技术应用中属于技术要求曾经比较高的，因此阅读本书读者最好具有一定的Java基础，HTML基础知识，MySQL数据库基础和Java Web应用开发等基础。

本书设计思路

本书结合当前技术发展趋势，以实用性为原则，由三个框架技术（Spring，Spring MVC，MyBatis）构建了本书的知识体系，重点讲解SSM框架在企业开发中常用的核心技术。内容逐层深入，先逐一讲解Spring，Spring MVC，MyBatis三大框架的精髓内容，再利用经典案例说明和实践。为保证读者学习效果，本书最后章节使用SSM框架技术搭建一个小而精的经典项目案例，通过此项目的实现可以加深读者对SSM框架技术的理解和掌握程度。

寄语读者

亲爱的读者朋友，感谢您在茫茫书海中找到并选择了本书。您手中的这本书，不是出自某知名出版社，更不是出自某位名师、大家。它的作者就在您的身边，希望它能够架起你我之间的学习、友谊的桥梁，希望它能带你轻松步入妙趣横生的编程世界，希望它会成为您进入IT编程行业的奠基石。

Java技术是无数人经验的积累，希望通过这本书的学习，能够从一些实例中领悟Java开发的精髓，并能够在合适的项目场景下应用它们。有了这本书做参考，将使得您在学习的过程中得到更多的乐趣。

本书可作为高等院校软件工程专业、计算机专业及相关专业本、专科学生的教材和参考书，亦适合于工程技术人员和程序设计人员参考。本书由任淑霞主编，张建军和陈香凝任副主编。由于时间仓促、学识有限，难免有不足和疏漏之处，恳请广大读者将意见和建议通过出版社反馈给我们，以便在后续版本中不断改进和完善。

编者

2018年6月

目录
Contents

第 1 篇　Spring 框架

学习目标：

- ☐ 熟悉 Spring IoC 管理依赖关系
- ☐ 熟悉 Spring AOP 面向切面编程
- ☐ 了解 Spring JDBC 数据库操作
- ☐ 熟悉 Spring 实现声明式事务和编程式事务
- ☐ 熟悉 XML 方式和全注解方式配置 IoC
- ☐ 精通 Spring 与 Spring MVC 框架集成开发
- ☐ 精通 Spring 与 MyBatis 框架集成开发
- ☐ 熟悉 SSM 集成框架开发企业应用

第 1 章　Spring 快速入手

本章要点（学会后请在方框中打钩）：

- □ 了解 Spring 框架
- □ 了解 Spring 框架的特点
- □ 掌握使用 Maven 搭建 Spring 项目的方法
- □ 编写 Spring 配置文件
- □ 了解 Bean 组件的定义
- □ 了解 Spring IoC/DI 注入思想和原理

1.1 Spring 的简介

Spring 是一个开源框架，是于 2003 年兴起的一个轻量级的 Java 开发框架，由 Rod Johnson 创建。简单来说，Spring 是一个分层的 JavaSE/EE full-stack(一站式)轻量级开源框架。

Spring 是一个轻量级框架，是当前主流的框架之一，它的目标是使现有技术更加易用，推进编码的最佳实践。

1.1.1 Spring 之父(图 1)

Rod Johnson 是 Spring Framework 的创始人，interface21 的 CEO

拥有丰富的 c/c++ 背景，丰富的金融行业背景

1996 年开始关注 Java 服务器端技术

为 Servlet2.4 和 JDO2.0 专家组成员

2002 年著《Expoert one-on-one J2EE 设计与开发》，改变了 Java 世界

技术主张是实用为本

音乐学博士

图 1　Rod Johnson

1.1.2 Spring 的特点

1. 轻量级(Lightweight)

Spring 核心包的容量不到 1MB，因此可以在很多小型设备中使用。

2. 非侵入性(No intrusive)

非侵入性增强应用程序组件的可重用性，减少对框架的依赖。

3. 容器(Container)

根据配置文件自动生成对象及属性等，不用编写任何代码来产生对象。

4.Inversion of Control(IoC)与 Dependency Injection(DI)

IoC 的目的是依赖于抽象；对象之间的关系由容器根据配置文件将依赖注入指定的对象中。

5.AOP(Aspect-oriented programming)

基于代理及拦截器的机制，与 Spring IoC 结合，运行时采用 Weaving 方式在 Spring 框

架的应用程序中使用各种声明式系统级服务。

6. 持久层（Persistent）

Spring 提供 DAO、编程事务与声明式事务，对于 ORM 工具（Hibernate、MyBatis）的整合及使用则可以进行简化。

7.Java EE 服务

Spring 可以使用 Java EE 标准规范提供的各项服务，并能与其无缝结合。

1.2 Spring 框架的优点

1. 使 Java EE 易用并能养成好良好的编程习惯

Spring 不重新开发已有的东西，因此 Spring 中没有日志记录的包，没有连接池，没有分布事务调度。这些服务由开源项目（例如 Commons Logging 用来作所有的日志输出，Commons DBCP 用来作数据连接池）提供，或由应用程序服务器提供。因为同样的原因，没有提供 O/R mapping 层，而是由 Hibernate 和 JPA 提供此服务。

2. 使已存在的技术更加易用

例如，尽管 Spring 没有底层事务协调处理工具，但提供了一个抽象层覆盖了 JTA 或任何其他的事务策略。

3. 没有直接和其他开源项目竞争

许多开发人员从来没有为 Struts 高兴过，并且感到在 MVC 框架中它还有改进的余地。在某些领域，轻量级的 IoC 容器和 AOP 框架与 Spring 有直接的竞争，但是在这些领域还没有较为流行的解决方案（Spring 在这些区域是开路先锋）。

4.Spring 在应用服务器之间是可移植的

保证可移植性一直是一项挑战，但是 Spring 避免任何特定平台或非标准化，并且支持在 WebLogic，Tomcat，Resin，JBoss，WebSphere 和其他的应用服务器上的用户。

5. 方便解耦，简化开发

Spring 提供的 IoC 容器可以将对象之间的依赖关系交由 Spring 进行控制，避免硬编码所造成的过度程序耦合。有了 Spring，用户便不必再为单实例模式类、属性文件解析等这些很底层的需求编写代码，而是专注于上层的应用。

6. 支持 AOP 编程

Spring 提供的 AOP 功能可以方便地进行面向切面的编程，许多不容易用传统 OOP 实现的功能都可以通过 AOP 轻松应付。

7. 支持声明式事务

在 Spring 中，开发人员可以摆脱单调烦闷的事务管理代码，通过声明的方式灵活地进行事务的管理，提高开发效率和质量。

8. 方便程序的测试

Spring 可以用非容器依赖的编程方式进行几乎所有的测试工作，在 Spring 里，测试不再是昂贵的操作，而是随手可做的事情。

9. 方便集成各种优秀框架

Spring 不排斥各种优秀的开源框架，相反，它可以降低各种框架的使用难度，提供了对各种优秀框架（如 Struts、Hibernate 等）的直接支持。

10. 降低 Java EE API 的使用难度

Spring 对很多难用的 Java EE API（如 JDBC、JavaMail、远程调用等）提供了一个薄薄的封装层，通过 Spring 的简易封装，这些 Java EE API 的使用难度大为降低。

当然，Spring 也有一定的缺点。自我感觉是所有框架共有的，但 Spring 的开发对设计要求较高，集成测试麻烦，对框架有一定的依赖性。

1.3 Spring 框架的 7 个模块

Spring 是一个开源框架，是为了解决企业应用程序开发的复杂性而创建的。框架的主要优势之一就是分层架构，分层架构允许用户选择使用哪一个组件，同时为 Java EE 应用程序开发提供集成的框架。Spring 框架的功能可以用在任何 Java EE 服务器中，大多数功能也适用于不受管理的环境。Spring 的核心要点是：支持不绑定到特定 Java EE 服务的可重用业务和数据访问对象。这样的对象可以在不同的 Java EE 环境（Web 或 EJB）、独立应用程序、测试环境之间重用。

Spring 框架是一个分层架构，由 7 个定义良好的模块组成。Spring 模块构建在核心容器之上，核心容器定义了创建、配置和管理 Bean 的方式，如图 2 所示。

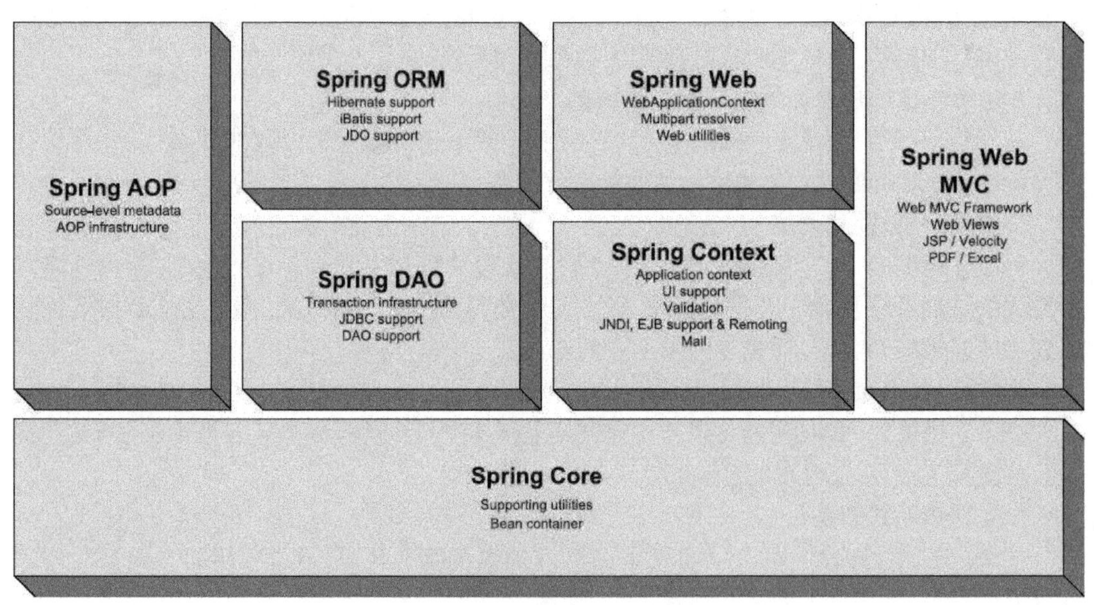

图 2　Spring 框架的 7 个模块

Spring 框架的 7 大核心模块内容如下。

（1）核心容器。核心容器提供 Spring 框架的基本功能。核心容器的主要组件是 Bean-

Factory，它是工厂模式的实现。BeanFactory 使用控制反转（IoC）模式将应用程序的配置和依赖性规范与实际的应用程序代码分开。

（2）Spring 上下文。Spring 上下文是一个配置文件，向 Spring 框架提供上下文信息。Spring 上下文包括企业服务，例如 JNDI 、EJB、电子邮件、国际化、校验和调度功能。

（3）Spring AOP。通过配置管理特性，Spring AOP 模块直接将面向切面的编程功能集成到 Spring 框架中。所以可以很容易地使 Spring 框架管理的任何对象都支持 AOP 。Spring AOP 模块为基于 Spring 的应用程序中的对象提供了事务管理服务。使用 Spring AOP，不依赖 EJB 组件，就可以将声明性事务管理集成到应用程序中。

（4）Spring DAO。JDBC DAO 抽象层提供了有意义的异常层次结构，可用该结构来管理异常处理和不同数据库供应商抛出的错误消息。异常层次结构简化了错误处理，并极大地降低了需要编写的异常代码数量（例如打开和关闭连接）。Spring DAO 的面向 JDBC 的异常遵从通用的 DAO 异常层次结构。

（5）Spring ORM。Spring 框架插入了若干个 ORM 框架，从而提供 ORM 的对象关系工具，其中包括 JDO 、Hibernate 和 iBatis SQL Map 。所有这些都遵从 Spring 的通用事务和 DAO 异常层次结构。

（6）Spring Web 模块。Web 上下文模块建立在应用程序上下文模块之上，为基于 Web 的应用程序提供了上下文。所以，Spring 框架支持与 Jakarta Struts 的集成。Web 模块还简化了处理多部分请求以及将请求参数绑定到域对象的工作。

（7）Spring MVC 框架。MVC 框架是一个全功能的构建 Web 应用程序的 MVC 实现。通过策略接口，MVC 框架成为高度可配置的，它容纳了大量视图技术，其中包括 JSP、Velocity、Tiles、iText 和 POI。

1.4 综合实例演示

编写 HelloSpring 类，输出“Hello, Spring DI !”

要求：“Spring DI”通过 Spring 注入 HelloSpring 类中。

实现步骤如下。

（1）添加 Spring 到项目中。

（2）编写配置文件。

（3）编写代码获取 HelloSpring 实例。

1.4.1 创建 Maven 工程

（1）打开 Eclipse，选择文件→新建→其他，其后操作如图 3、图 4 所示。

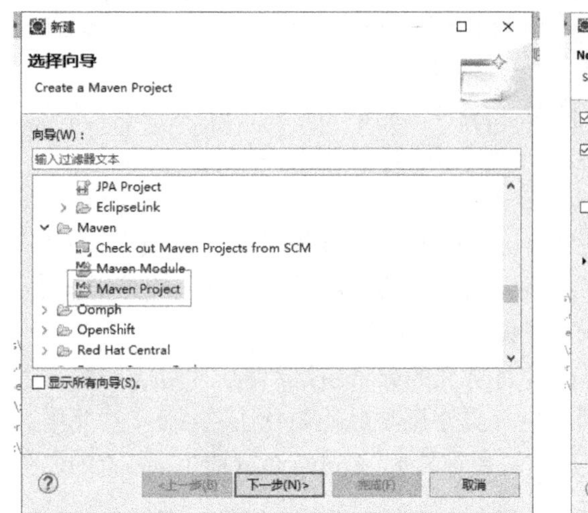

图 3　选择“Maven Project”

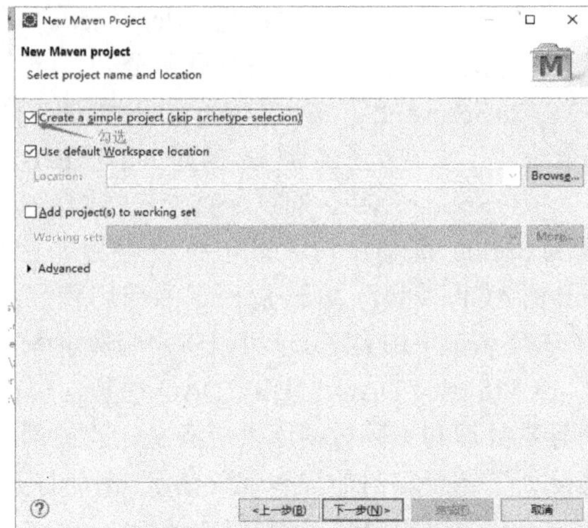

图 4　勾选选项

（2）填写 Group Id 和 Artifact Id，Version 默认为“0.0.1-SNAPSHOT”，Packaging 默认为“jar”，Name 与 Description 选填，其他内容可不填，如图 5 所示。

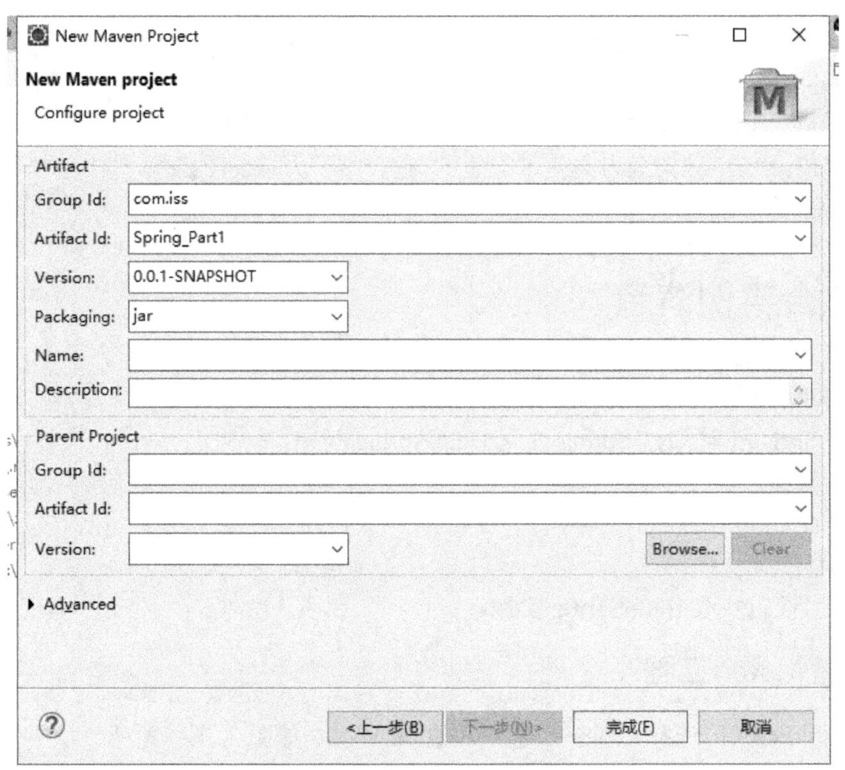

图 5　新建 Maven 项目界面

1.4.2　项目工程文件结构

项目工程的文件结构如图 6 所示。

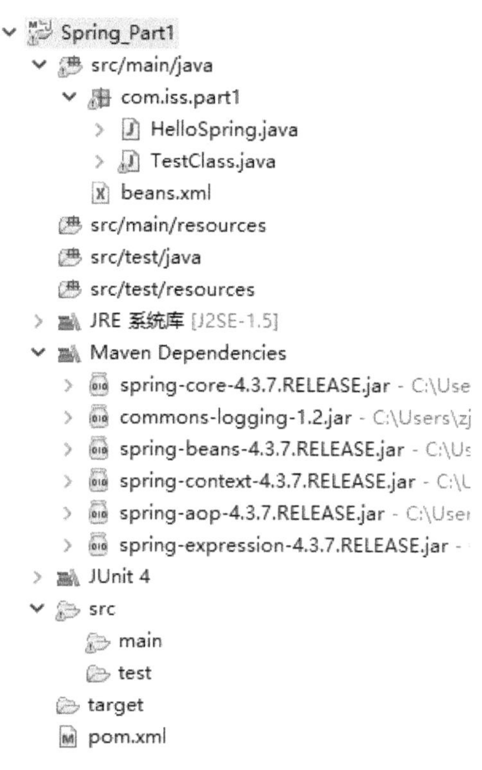

图 6　项目工程文件结构

1.4.3　修改 pom.xml 文件添加依赖

修改 pom.xml 文件添加依赖，如图 7 所示。

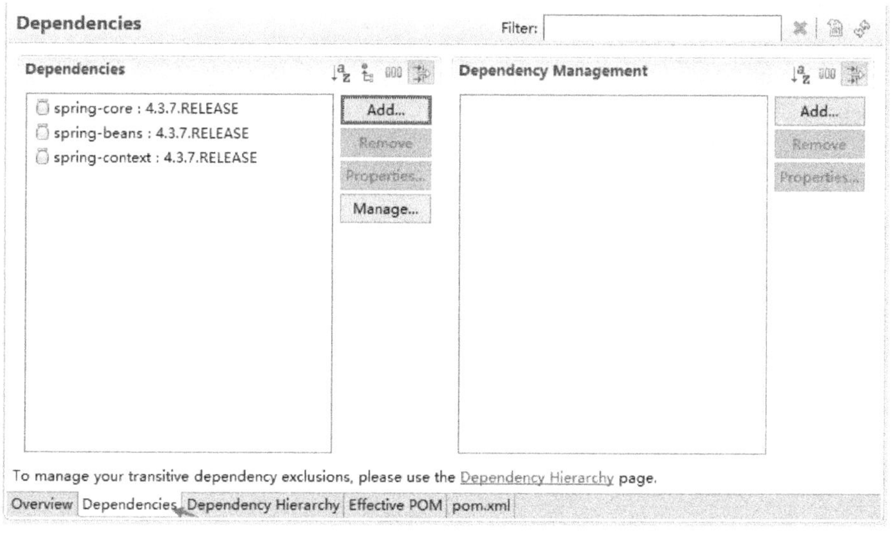

图 7　pom.xml 添加依赖

pom.xml 的代码内容如下。

```
<project xmlns="http://maven.apache.org/POM/4.0.0"
xmlns:xsi="http://www.w3.org/2001/XMLSchema-instance"
xsi:schemaLocation="http://maven.apache.org/POM/4.0.0
http://maven.apache.org/xsd/maven-4.0.0.xsd">
    <modelVersion>4.0.0</modelVersion>
    <groupId>com.iss</groupId>
    <artifactId>ProjectX</artifactId>
    <version>0.0.1-SNAPSHOT</version>
    <name>Project</name>

    <dependencies>
    <dependency>
    <groupId>org.springframework</groupId>
    <artifactId>spring-core</artifactId>
    <version>4.3.7.RELEASE</version>
    </dependency>
    <dependency>
    <groupId>org.springframework</groupId>
    <artifactId>spring-context</artifactId>
    <version>4.3.7.RELEASE</version>
    </dependency>
    </dependencies>
</project>
```

1.4.4 创建 HelloSpring.java 文件

创建 HelloSpring.java 文件的代码如下。

```
// 用来作为我们演示业务层的功能:
public class HelloSpring {
public void sayHelloSpring(String str) {
System.out.println("Hello " + str);
}
}
```

1.4.5 创建 TestClass.java 文件

创建 TestClass.java 文件的代码如下。

```
// 用来作为我们演示控制层的功能：
public class TestClass {
@Test
public void testSayHello( ) {
// 首先读取配置文件，配置文件中的 bean 将会保存到 ApplicationContext 的实例中
ApplicationContext ac = new ClassPathXmlApplicationContext("classpath*:beans.xml");

// 从 ApplicationContext 的实例中按照 id 值获取对应类的实例对象，并且需要进行强制类型
HelloSpring hs = (HelloSpring) ac.getBean("helloworld");
 // 使用对象的内部方法，就像我们使用 new 创建的对象一样
 hs.sayHelloSpring("Spring DI !");
 }
 }
```

1.4.6 创建 beans.xml 文件

创建 beans.xml 文件的代码如下。

```
//spring 的核心配置文件，注意命名空间的使用
<?xml version="1.0" encoding="UTF-8"?>
<beans xmlns="http://www.springframework.org/schema/beans"
xmlns:xsi="http://www.w3.org/2001/XMLSchema-instance"
xmlns:context="http://www.springframework.org/schema/context"
xsi:schemaLocation="http://www.springframework.org/schema/beans
 http://www.springframework.org/schema/beans/spring-beans-4.0.xsd
 http://www.springframework.org/schema/context
 http://www.springframework.org/schema/context/spring-context-4.0.xsd">
<bean id="helloworld" class="com.iss.part1.HelloSpring"></bean>
</beans>
```

1.4.7 测试运行输出结果

测试运行输出结果如图 8 所示。

图 8　输出结果

专家讲解

1. <?xml version="1.0" encoding="UTF-8"?> 表示这是 xml 文件的开头声明，xml 使用的版本以及编码方式，一般为通用配置。

2. xmlns 表示 XML Namespaces 的缩写，中文名称是 XML 命名空间。这里使用的默认的命名空间：xmlns="http://www.springframework.org/schema/beans"。在 xml 文件中，标签都是我们自定义的，但是我们自定义的标签可能与别人定义的产生冲突，所以，这里我们需要使用一个命名空间来标识。可以将命名空间理解成一个包的名称，包的名称是自定义的。包内部的方法、参数可能与别的包内部的重名，但是在两个包里面，就不需要担心冲突了。

3. xmlns:xsi 表示 xml 文件遵守了 xsi 所定义的规范。这里 xsi 的全称是 xml schema instance。它代替了后面的 url 地址，即“http://www.w3.org/2001/XMLSchema-instance”，需要特别说明的是这里的 xsi 只是人们习惯上定义的名称，我们也可以换一个其他的名称，不过无论怎么取，它都是 url 地址的别名。实际解析 xml 文档时，解析器就会按照 url 的地址规定的规则来解释当前的 xml 文档。

4. xsi:schemaLocation 表示 xml 书写遵循的语法。schemaLocation 是其一个属性，该属性由两部分组成：前半部分是命名空间的名称，后半部分是 xsd 的地址。它的作用是把这个地址表示的 xml 文件引进来，方便 eclipse 等开发工具判断 xml 是否符合语法。这里的作用等价于 <import namespace="xxx" schemaLocation="xxx.xsd"/>。

5. bean 表示我们让 spring 托管的一个对象，这个对象的 id 是 helloworld，id 是 class 对象的别名，我们使用时调用这个 id 就表示调用后面的 class 表示的对象。需要特别注意的是，这个 id 是唯一的，不能够重复的。

小结

（1）Spring 的核心是一个轻量级（Lightweight）的容器（Container）。

（2）Spring 是实现 IoC（Inversion of Control）容器和非入侵性（No intrusive）的框架。

（3）Spring 提供 AOP（Aspect-oriented programming）概念的实现方式。

（4）Spring 提供对持久层（Persistence）、事物（Transcation）的支持。

（5）Spring 提供 MVC Web 框架的实现，并对一些常用的企业服务 API（Application

Interface) 提供一致的模型封装。

(6) Spring 提供了对现存的各种框架（Struts、JSF、Hibernate、MyBatis、Webwork 等）相整合的方案。

总之，Spring 是一个全方位的应用程序框架。Spring 框架功能是非常强大的，单独使用可能感觉不是很深，和其他的框架结合使用，就会彰显出它的魅力了。

经典面试题

(1) 什么是 Spring?
(2) Spring 框架有哪些优点？
(3) Spring 框架都能帮我们做什么？
(4) Spring 框架的目标是什么？
(5) Spring 框架有哪些模块？
(6) 如何编写 Spring 核心配置文件？
(7) Spring 当前最新版本号是多少？
(8) spring 与 struts 有什么区别？
(9) 如何解释 Spring 的核心容器模块？
(10) 如何解释 Spring 的 AOP 模块？

跟我上机

(1) 使用 Eclipse 创建 Maven 项目，使用 Spring 依赖注入方式完成图 9 所示功能输出。

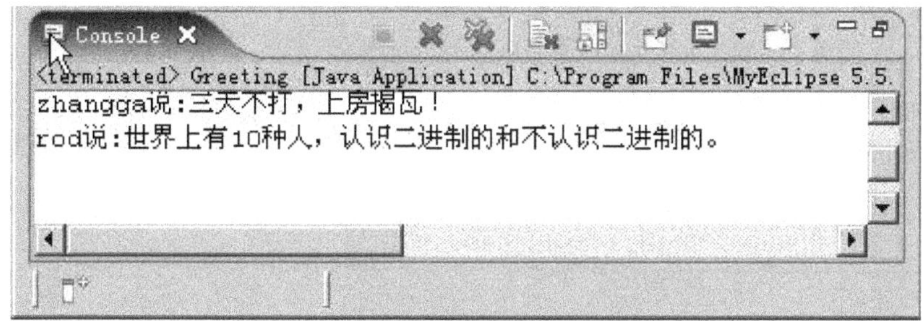

图 9　功能输出

要求：说话人和说话内容都通过 Spring 注入。
实现思路：
①添加 Spring 到项目；
②编写程序代码和配置文件（同时配两个 Bean）；
③获取 Bean 实例，测试功能方法。

第 2 章　Spring IoC(控制反转) /DI(依赖注入)

本章要点(学会后请在方框中打钩):

- □ 了解什么是 IoC 和 DI
- □ 掌握使用多种方式实现依赖注入
- □ 理解自动装配
- □ 掌握方法的注入与方法替换
- □ 理解 Bean 之间的关系
- □ 理解 Bean 的作用域
- □ 掌握 Spring 配置文件拆分策略

2.1 Spring IoC/DI 介绍

Spring 的 IoC（Inversion of Control，控制反转）是一个非常重要的概念，对于很多人来说即使看过很多资料，做过很多实例但仍会不知所措。

Spring DI（Dependency Injection，依赖注入），在 IoC 容器运行期间，动态地将某种依赖关系注入对象之中。

IoC 和 DI 有什么关系呢？其实它们是同一个概念不同角度的描述，由于控制反转概念比较含糊（可能只是理解为容器控制对象这一个层面，很难让人想到谁来维护对象关系），所以 2004 年大师级人物 Martin Fowler 又给出了一个新的名字："依赖注入"，相对 IoC 而言，"依赖注入"明确描述了"被注入对象依赖 IoC 容器配置依赖对象"。

专家讲解

控制反转 IoC 和依赖注入 DI 是同一个概念，因为翻译的不同，而出现的两个名字。

1. 控制反转的意思就是说，当我们调用一个方法或者类时，不再由我们主动去创建这个类的对象，而是把控制权交给别人（Spring）。

2. 依赖注入的意思就是说，Spring 主动创建被调用类的对象，然后把这个对象注入我们自己的类中，使得我们可以使用它。

专家举例

程序员加了一个月班，很累，想要放松一下，于是想找人一起去吃麻辣烫。

不使用 Spring 的传统做法是，我们自己通过微信等软件，主动寻找目标，花费大量人力物力，达成协议后，申请"场所"办正事。

而使用 Spring 的做法就很方便了，我们直接去某个场所，那个地方直接就有目标等候。

2.2 Spring IoC 实现

专家举例

"自由恋爱"和"包办婚姻"的例子。

假设：有一个男孩叫小明，有一个事件叫结婚，有多个女孩分别叫小美、小丽等。

在自由恋爱时，即当没有使用控制反转时：小明想要结婚，那么小明首先需要寻找一个合适的对象，再针对这个对象进行各种形式的追求，最后才能结婚。结婚这个动作从开始到结束一直"强制"绑定了双方，在代码中是一种紧耦合的表现。在这个过程里，小明为了结婚，付出了很大的"牺牲"，如果运气不好，再来一次的话，那么从寻找对象到结婚这个动作就必须重来。

在包办婚姻情况下，即当使用控制反转时：小明想要结婚，那么 Spring 在结婚双方之间就像婚介所（也可以换成父母），它管理所有女孩的信息，小明只要说：我想找个上得厅堂，下得厨房的女孩，那么“婚介所”就会从多个女孩中寻找到符合要求的人提供给小明，他们直接结婚即可。如果小明在结婚时问“是否愿意嫁给我”，女孩回答“不愿意”的话，那么程序就直接报错。因此，小明可以看作一直“站在民政局门口”，只要成功取得一个对象，那么马上就能完成“结婚”这个动作。

下面我们先来看看“自由恋爱”情况下的“结婚”动作发生的全过程，使用 3 种方式完成对比。

2.2.1 使用 Java 面向对象思想实现

1）创建 Maven 项目 Spring_Part2

结构图略（参照第 1 章）。

2）创建 Boy.java

```
package com.iss.demo;
import org.junit.Test;
public class Boy {
@Test
public void testMarry( ) {
  Married getMarried = new Married( );
  getMarried.getMarried( );
}
}
```

3）创建 Married.java

```
package com.iss.demo;
public class Married {
public void getMarried( ) {
  XiaoMei xm = new XiaoMei( );
  /* 对小美进行如下动作 */
  // 追求女…
  // 旅游…
  // 看电影…
  // 玫瑰花…
  // 逗开心…
  xm.sayYes( );
}
```

4）创建 XiaoMei.java

```
public class XiaoMei {
public void sayYes( ) {
  System.out.println(" 小美 :" + " 我愿意 !");
}
}
```

5）创建 XiaoLi.java

```
public class XiaoLi {
public void sayYes( ) {
System.out.println(" 小丽 :" + " 我愿意 !");
}
}
```

6）测试运行输出结果

测试运行输出结果如图 1 所示。

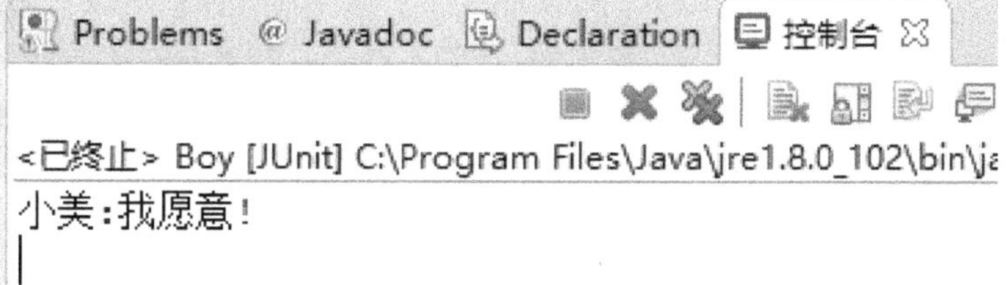

图 1　测试运行输出结果

专家讲解

当 Boy 调用 Married 时，我们需要创建 XiaoMei 对象，并且需要针对当前的 XiaoMei 对象进行一系列的动作处理，最后 XiaoMei 对象才能被我们使用。当我们创建的对象变为 XiaoLi 时，我们需要修改 Married 中的全部内容，才能使用 XiaoLi 对象。这么做的缺点显而易见，因此，我们需要换一种方式来改造我们的 getMarried 方法，使得我们在改变创建的对象时，getMarried 方法整体上不用修改。

2.2.2　使用 Java 多态性思想改造

1）创建 Girl.java 文件

```
public interface Girl {
public void sayYes( );
}
```

2）修改 XiaoMei.java，XiaoLi.java 实现此接口

```
package com.iss.demo;
public class XiaoMei implements Girl {
public void sayYes( ) {
  System.out.println(" 小美 :" + " 我愿意 !");
}
}
```

3）修改 Married.java 文件

```
package com.iss.demo;
public class Married {
Girl girl;
public Girl getGirl( ) {
  return girl;
}
public void setGirl(Girl girl) {
  this.girl = girl;
}
public void getMarried( ) {
  girl.sayYes( );
}
}
```

4）修改 Boy.java

```
package com.iss.demo;
import org.junit.Test;
public class Boy {
@Test
public void testMarry( ) {
  Married getMarried = new Married( );
  getMarried.setGril(new XiaoMei( ));
  getMarried.getMarried( );
}
}
```

5）测试运行输出结果

测试运行输出结果如图 2 所示。

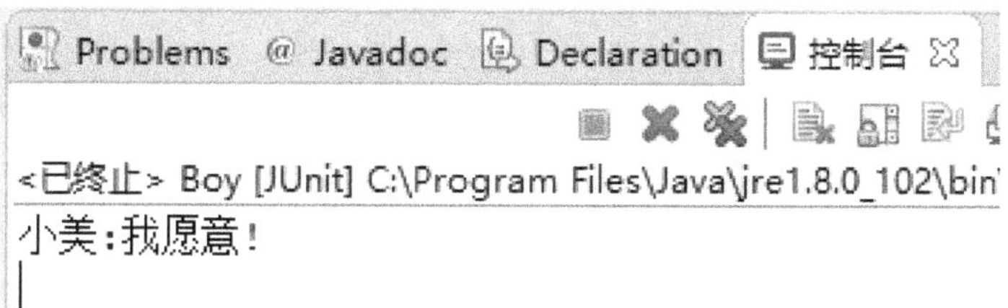

图 2　测试运行输出结果

专家讲解

当 Boy 想要使用 getMarried 时，就不需要在 getMarried 中实现创建对象等动作，而是直接使用一个符合要求的对象即可。

2.2.3　使用 Spring 来改造

1) 创建 beans.xml

注：存放位置为 src/main/resources，具体代码内容如下。

```
// 省略 Beans 和命名空间
<bean id="XiaoLi" class="com.iss.demo.XiaoLi"></bean>
<bean id="XiaoMei" class="com.iss.demo.XiaoMei"></bean>
<bean id="GetMarried" class="com.iss.demo.Married">
        <property name="gril" ref="XiaoLi"></property>
</bean>
```

专家讲解

上面的配置文件中，我们首先针对每一个对象创建一个 bean，再针对结婚这个动作创建一个 bean，并且结婚的 bean 的内部有一个结婚对象的属性，属性名称对应 Married.java 中的属性名称，属性的值使用 ref 设置为另外两个 bean 的 id 值。关于这个属性的设置及用法，我们将在后续文章中进行介绍。

2）修改 Boy.java

```
package com.iss.demo;
// 省略导入类
public class Boy {
@Test
public void testMarry( ) {
        ApplicationContext ac = new ClassPathXmlApplicationContext("beans.xml");
```

```
        Married gm = (Married) ac.getBean("GetMarried");
        gm.getMarried( );
    }
}
```

3）运行输出结果

运行输出结果如图 3 所示。

```
Problems  @ Javadoc  Declaration  控制台
<已终止> Boy [JUnit] C:\Program Files\Java\jre1.8.0_102\bin\ja
六月 03, 2017 11:57:48 下午 org.springframework
信息: Refreshing org.springframework.context.s
六月 03, 2017 11:57:49 下午 org.springframework
信息: Loading XML bean definitions from class
小丽:我愿意!
```

图 3　测试运行输出结果

专家讲解

从 test 方法中，我们可以看到，只保留了结婚的动作，彻底隐藏了结婚对象，结婚对象交给 Spring 去管理。当结婚对象发生改变时，我们修改的是 Spring 配置文件，而不是源代码。

2.3　Spring DI（依赖注入）

1）Spring bean 的概念

Spring 等价于一个对象的托管工厂，Spring bean 就是托管工厂里组装生产出来的实例对象。每一个 bean 对应一个 id，一个 bean 表示一个类的实体对象。

2）依赖注入（DI）的概念

我们可以把这个高大上的“依赖注入”理解为 Spring 帮助我们实例化一个对象的过程。再通俗一点来说就是 new 对象这个动作交由 Spring 托管工厂完成。

依赖注入的方法，也就是 Spring 支持的实例化对象的方法，包括：属性注入、构造函数注入、索引注入、工厂方法注入、泛型注入等。

2.3.1 属性注入

1）创建 Customer.java 文件

```
package com.iss.demo2;
public class Customer {
    private String name;
    private String sex;
    private int age;
       //geter 和 seter 省略

    // 此空参数构造函数一定要有
    public Customer( ) {
    }
    public Customer(String name, String sex, int age) {
        super( );
        this.name = name;
        this.sex = sex;
        this.age = age;
    }
    @Override
     public String toString( ) {
        return "Customer [name=" + name + ", sex=" + sex + ", age=" + age + "]";
     }
}
```

2）创建 beans.xml 文件

注：这里 property 中的 name 的值与实际对象中的属性值一一对应。

```
<bean id="Customer1" class="com.iss.demo2.Customer"></bean>
<bean id="Customer2" class="com.iss.demo2.Customer">
 <property name="name" value="Tom"></property>
 <property name="sex" value="male"></property>
 <property name="age" value="22"></property>
</bean>
```

3）编写测试类

```
ApplicationContext ac = new ClassPathXmlApplicationContext("beans.xml");
Customer customer2 = (Customer) ac.getBean("Customer2");
System.out.println(customer2);
```

4）测试运行输出结果

测试运行输出结果如图 4 所示。

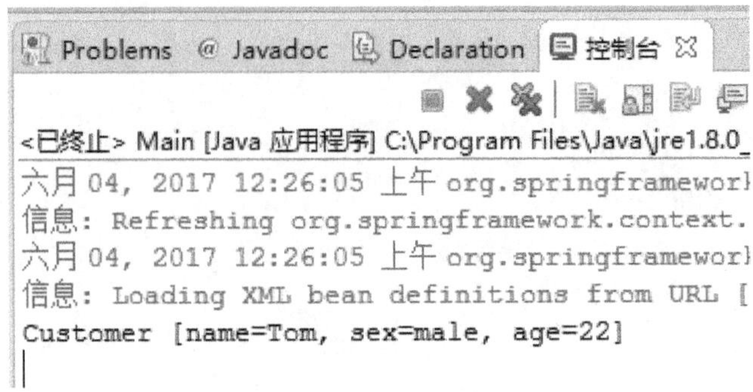

图 4　测试运行输出结果

2.3.2　构造函数注入

1）修改 beans.xml 文件

```
<bean id="Customer3" class="com.iss.demo2.Customer">
<constructor-arg type="String" value="Jack"> </constructor-arg>
<constructor-arg type="String" value="male"> </constructor-arg>
<constructor-arg type="int" value="23"> </constructor-arg>
</bean>
```

2）测试运行输出结果

测试运行输出结果如图 5 所示。

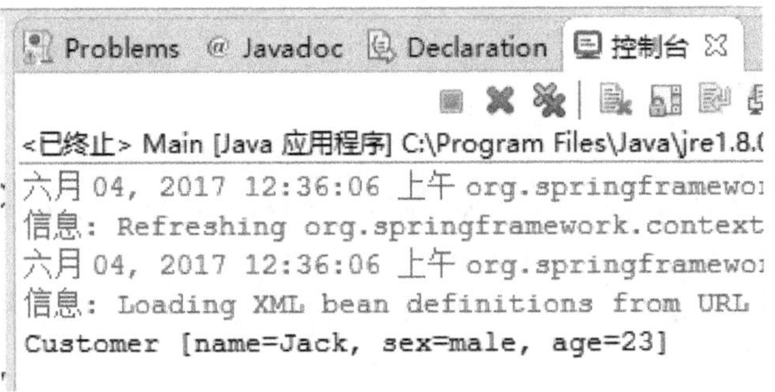

图 5　测试运行输出结果

专家讲解

这里 value 值是有顺序的，即构造函数参数的顺序要与这里一对一，如果错位，如 Jack 与 male 换位置，大家可以实际运行一下并观察结果。

2.3.3 索引注入

1）修改 beans.xml 文件

```
<bean id="Customer4" class="com.iss.demo2.Customer">
    <constructor-arg index="0" value="Emma"> </constructor-arg>
    <constructor-arg index="1" value="female"> </constructor-arg>
    <constructor-arg index="2" value="25"> </constructor-arg>
</bean>
```

2）测试运行输出结果

测试运行输出结果如图 6 所示。

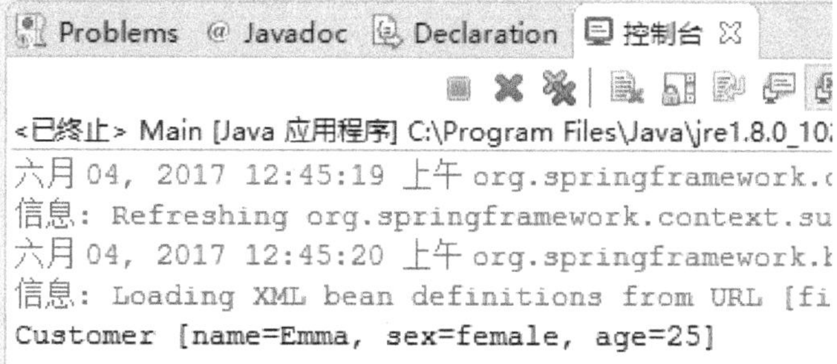

图 6 测试运行输出结果

专家讲解

这里 value 值最好与构造函数中声明的属性值的类型保持一致，防止发生类型转化错误。我们可以联合使用构造函数注入和索引注入两种方式。同时还解决了构造函数中属性值必须按照顺序来书写的问题。

```
1. <bean id="Customer5" class="com.iss.demo2.Customer">
2.   <constructor-arg index="0"  type="String" value="Emma"></constructor-arg>
3.   <constructor-arg index="1"  type="String" value="female"></constructor-arg>
4. <constructor-arg index="2"  type="int" value="25"></constructor-arg>
5. </bean>
```

在联合使用的方式下，就不必担心构造函数值的顺序问题，如这里我们把 Emma 与 female 的两行（行 2，行 3）的顺序进行调换，其输出结果仍然是一致的。

2.3.4 工厂方法注入（非静态）

1）修改 beans.xml 文件

```
<bean id="CustomerFactory" class=" com.iss.demo2.CustomerFactory"></bean>
<bean id="customer" factory-bean="CustomerFactory" factory-method="createCustomer"/>
```

2）创建 CustomerFactory.java 的 createCustomer 方法

```
package com.iss.demo2;
public class CustomerFactory {
  public Customer createCustomer( ) {
        Customer customer = new Customer( );
        customer.setName("Ingo");
        customer.setAge("26");
        customer.setSex("male");
        return customer;
  }
}
```

3）测试运行输出结果

测试运行输出结果如图 7 所示。

图 7　测试运行输出结果

2.3.5　工厂方法注入（静态）

1）修改 beans.xml 文件

```
<bean id="Customer"
class="com.iss.demo2.StaticCustomerFactory"    factory-method="createCustomer"></
bean>
```

2）创建 StaticCustomerFactory.java

```
package com.iss.demo2;
  public class StaticCustomerFactory {
    public static Customer createCustomer( ){
        Customer customer = new Customer( );
        customer.setName("Rose");
        customer.setAge(27);
        customer.setSex("male");
```

```
            return customer;
        }
```

3）测试运行输出结果

测试运行输出结果如图 8 所示。

图 8　测试运行输出结果

2.3.6　bean 方式注入

1）修改 beans.xml 文件

```
    <bean id="Record2" class="com.iss.demo2.Record">
            <property name="company" value="ABCD"></property>
            <property name="position" value="Engineer"></property>
            <property name="address" value="Beijing"></property>
    </bean>
    <bean id="Customer2" class="com.iss.demo2.Customer">
            <property name="name" value="Tom"></property>
            <property name="sex" value="male"></property>
            <property name="age" value="22"></property>
            <property name="record" ref="Record2"></property>
    </bean>
```

2）创建 Record.java 文件

```
package com.iss.demo2;
public class Record {
    private String company;
    private String position;
    private String address;
    // 由于篇幅关系，请读者自行添加 set/get 方法，构造函数，重写 toString 方法等
}
```

3）修改 Customer.java

```
public class Customer {
private String name;
private String sex;
private int age;
private Record record;
// 省略 setter 和 getter
@Override
public String toString( ) {
        return "Customer [name=" + name + ", sex=" + sex + ", age=" + age + "],Record
[company=" + record.getCompany( )+ ",address=" + record.getAddress( ) + "]";
}
```

4）测试运行输出结果

测试运行输出结果如图 9 所示。

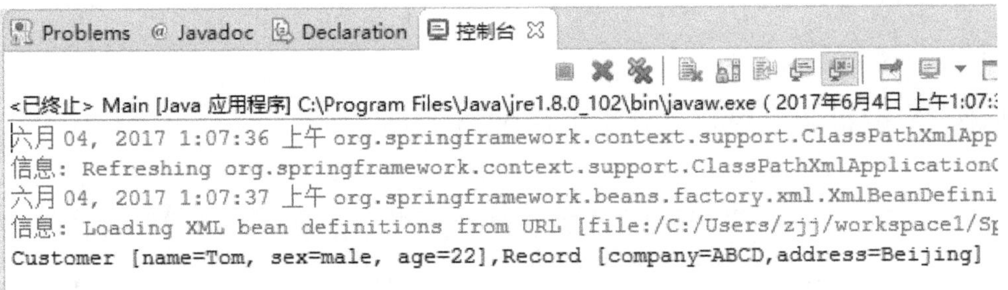

图 9　测试运行输出结果

2.3.7　使用注入内部 bean 的方式注入

1）修改 beans.xml 文件

```
<bean id="Customer3" class="com.iss.demo2.Customer">
            <property name="name" value="Tom"></property>
            <property name="sex" value="male"></property>
            <property name="age" value="22"></property>
            <property name="record">
                  <bean class="com.iss.demo2.Record">
                        <property name="company" value="ABCD"></property>
                        <property name="position" value="Engineer"></property>
                        <property name="address" value="Beijing"></property>
                  </bean>
```

```
            </property>
    </bean>
```

2）测试运行输出结果

测试运行输出结果如图 10 所示。

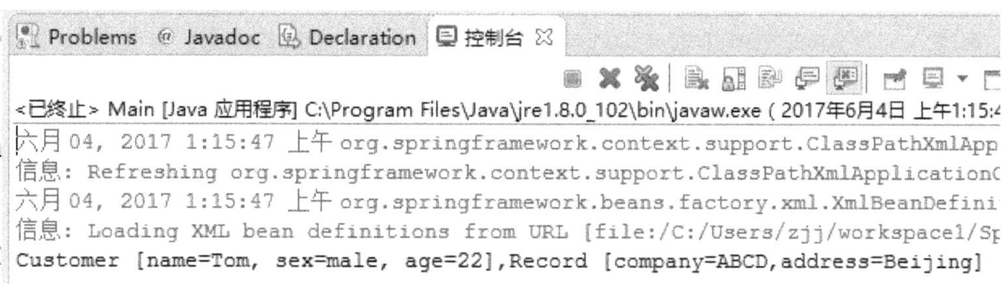

图 10　测试运行输出结果

2.3.8　使用空值的 bean 方式注入

```
// 举个例子，该客户还没有被记录到档案中，仅仅保存基本信息。那么 Record 就是空。
<bean id="Customer4" class="com.iss.demo2.Customer">
        <property name="name" value="Tom"></property>
        <property name="sex" value="male"></property>
        <property name="age" value="22"></property>
        <property name="record">
            <null></null>
        </property>
    </bean>
```

2.3.9　级联属性注入

```
<bean id="Customer5" class="com.iss.demo2.Customer">
        <property name="name" value="Tom"></property>
        <property name="sex" value="male"></property>
        <property name="age" value="22"></property>
        <property name="record.company" value="ABCD"></property>
    </bean>
```

专家提示

在 Customer.java 中，将 private Record record; 替换为 private Record record = new Record( );

2.3.10 集合属性注入

```
// 在 Customer.java 中添加属性 private List<String> hobbies = new ArrayList<String>( ); 并且生成对应的 set、get 方法，重写 toString 方法。
<bean id="Customer6" class="com.iss.demo2.Customer">
        <property name="name" value="Tom"></property>
        <property name="sex" value="male"></property>
        <property name="age" value="22"></property>
        <property name="record" ref="Record2"></property>
        <property name="hobbies">
            <list>
                <value>sing</value>
                <value>dance</value>
            </list>
        </property>
    </bean>
```

2.4 自动装配

2.4.1 什么是自动装配

Spring IoC 容器可以自动装配（autowire）相互协作 bean 之间的关联关系。因此，如果可能的话，可以让 Spring 通过检查 BeanFactory 中的内容，来替我们指定 bean 的协作者（其他被依赖的 bean）。自动装配一共有六种类型，如表 1 所示。由于自动装配可以针对单个 bean 进行设置，因此可以让有些 bean 使用 autowire，有些 bean 不使用。自动装配的方便之处在于能减少或者消除属性或构造器参数的设置，从而可以让配置文件缩减。在 xml 配置文件中，可以在 <bean/> 元素中使用 autowire 属性指定。

表 1 自动装配的六种类型

模　式	说　明
Default	在每个 bean 中都有一个 autowire=default 的默认配置，它的含义是：采用 beans 和跟标签中的 default-autowire=" 属性值 " 一样的设置
No	不使用自动装配，必须通过 ref 元素指定依赖，默认设置
ByName	根据属性名自动装配。此选项将检查容器并根据名字查找与属性完全一致的 bean，并将其与属性自动装配。例如，在 bean 定义中将 autowire 设置为 byName，而该 bean 包含 master 属性（同时提供 setMaster(...) 方法），Spring 就会查找名为 master 的 bean 定义，并用它来装配 master 属性

续表

模　式	说　明
ByType	如果容器中存在一个与指定属性类型相同的 bean，那么将与该属性自动装配。如果存在多个该类型的 bean，那么将会抛出异常，并指出不能使用 byType 方式进行自动装配。若没有找到相匹配的 bean，则什么事都不发生，属性也不会被设置。如果不希望这样，可以通过设置 dependency-check="objects" 让 Spring 抛出异常
Constructor	与 byType 的方式类似，不同之处在于它应用于构造器参数。如果在容器中没有找到与构造器参数类型一致的 bean，那么将会抛出异常
Antodetect	通过 bean 类的自省机制（introspection）来决定是使用 constructor 还是 byType 方式进行自动装配。如果发现默认的构造器，那么将使用 byType 方式

2.4.2　自动装配

专家讲解

autowire 属性可以设置为 no、byType 或 byName。若采用 byType 方式，当 Spring 无法决定注入哪个 Bean 的时候，将报错。

在 Spring 中使用自动装配来提供我们的 bean 有以下几种方式。

1）autowire=No

这是默认的模式，与使用 ref 将 Record 装配到 Customer 中相同。

2）autowire=byName

此种方式表示按照名称来装配，即 bean 的 id 值需要与 Customer 中的属性值保持一致。

专家举例

```
<bean id="Customer1" class="com.iss.demo2.Customer" autowire="byName">
    <property name="name" value="Tom"></property>
    <property name="sex" value="male"></property>
    <property name="age" value="22"></property>
  <!-- <property name="record" ref="Record2"></property>
      如果上面使用 autowire="byName" 注入可以注释掉这句 -->
</bean>
<bean id="record" class="com.iss.demo2.Record">
    <property name="company" value="ABCD"></property>
    <property name="position" value="Engineer"></property>
    <property name="address" value="Shanghai"></property>
</bean>
```

专家讲解

1. 这里的 autowire = byName 添加在 Customer 的 bean 之上，即外部的 bean 对象之上。

2.bean 的 id 值 record 必须完全等于 Customer 对象中的属性值。

3）autowire=byType

此种方式表示按照类型来装配，即 bean 的类型需要与 Customer 的属性值的 id 值保持一致。

此处代码将上面的 autowire=byName 修改为 autowire=byType 即可。

这里按照类型来注入时，bean 的配置中不能出现两个类型相同的 bean，即这里不能出现两个 record，即使这里的 id 值是不同的。

4）autowire="constructor"

此种方式表示按照构造函数来装配，即这里的 record 的对象类型与 Customer 中的构造函数的参数类型是一致的。

此处代码将上面的 autowire=byType 修改为 autowire="constructor"。

在 Customer.java 中，增加一个构造函数，具体内容如下。

```
public Customer(Record record) {
        super( );
        this.record = record;
    }
```

5）autowire=default

在每个 bean 中都有一个 autowire=default 的默认配置，它的含义是采用 beans 和跟标签中的 default-autowire=" 属性值 " 一样的设置。

修改 beans.xml，具体代码如下。

```
<?xml version="1.0" encoding="UTF-8"?>
<beans xmlns="http://www.springframework.org/schema/beans"
    xmlns:xsi="http://www.w3.org/2001/XMLSchema-instance"
    xsi:schemaLocation="http://www.springframework.org/schema/beans
        http://www.springframework.org/schema/beans/spring-beans.xsd"
    default-autowire="byName">
  <bean id="Customer1" class="com.iss.demo2.Customer"  autowire="default">
        <property name="name" value="Tom"></property>
        <property name="sex" value="male"></property>
        <property name="age" value="22"></property>
  </bean>
  <bean id="record" class="com.iss.demo2.Record">
```

```
            <property name="company" value="ABCD"></property>
            <property name="position" value="Engineer"></property>
            <property name="address" value="Beijing"></property>
    </bean>
</beans>
```

2.5 方法注入

1）修改单元测试方法

```
@org.junit.Test
    public void test1( ) {
        System.out.println("Start Test( )");
        // 省略获得 ac 对象的语句
        Record customer1 = (Record)ac.getBean("record2");
        Record customer2 = (Record)ac.getBean("record2");
        System.out.println(customer1==customer2);
    }
```

2）修改 beans.xml 配置文件

```
<bean id="Customer1" class="com.iss.demo2.Customer">
        <property name="name" value="Tom"></property>
        <property name="sex" value="male"></property>
        <property name="age" value="22"></property>
</bean>
<bean id="record" class="com.iss.demo2.Record">
        <property name="company" value="ABCD"></property>
        <property name="position" value="Engineer"></property>
        <property name="address" value="Beijing"></property>
</bean>
<bean id="record2" class="com.iss.demo2.Record">
        <property name="company" value="ABCD"></property>
        <property name="position" value="Engineer"></property>
        <property name="address" value="Beijing"></property>
</bean>
```

3）运行输出结果

运行输出结果如图 11 所示。

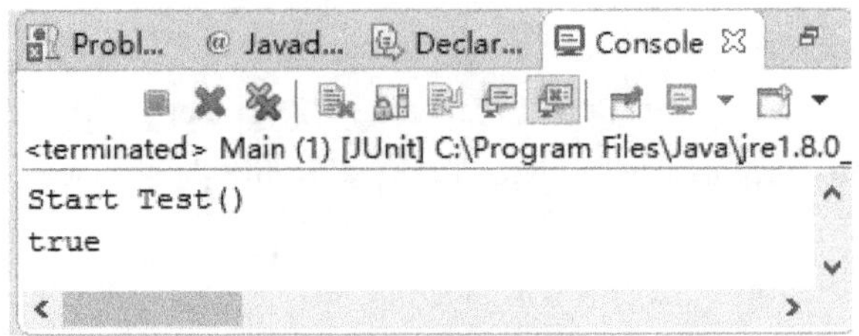

图 11　运行输出结果

专家讲解

Spring 所管理的 bean 是单例模式的，即前后两次取得的对象是完全相同的。

4）修改 id 为 record2 的 bean.xml

```
<bean id="record2" class="com.iss.demo2.Record" scope="prototype">
            <property name="company" value="ABCD"></property>
            <property name="position" value="Engineer"></property>
            <property name="address" value="Beijing"></property>
    </bean>
```

5）测试运行输出结果

测试运行输出结果如图 12 所示。

图 12　测试运行输出结果

专家讲解

输出 false，即两次取得的结果是不相同的。说明每次获取的 bean 都是新生成的。

专家讲解

prototype 作用域的 bean 会导致在每次对该 bean 请求（将其注入另一个 bean 中，或者以程序的方式调用容器的 getbean( ) 方法）时都会创建一个新的 bean 实例。根据经验，对有状态的 bean 应使用 prototype 作用域，而对无状态的 bean 则应该使用 singleton 作用域。

> 对于具有 prototype 作用域的 bean 来说，有一点很重要，即 Spring 不能对该 Bean 的整个生命周期负责。具有 prototype 作用域的 Bean 创建后交由调用者负责销毁对象回收资源。简单地说就是 singleton 只有一个实例，即单例模式。prototype 访问一次创建一个实例，相当于 new。

6）修改 id 为 Customer1 的 bean

```
<bean id="Customer1" class="com.iss.demo2.Customer" scope="prototype">
        <property name="name" value="Tom"></property>
        <property name="sex" value="male"></property>
        <property name="age" value="22"></property>
        <property name="record" ref="record"></property>
</bean>
```

7）修改测试方法

```
@org.junit.Test
        public void test1( ) {
                System.out.println("Start Test( )");
                Customer customer1 = (Customer)ac.getBean("Customer1");
                Customer customer2 = (Customer)ac.getBean("Customer1");
                System.out.println(customer1==customer2);
                System.out.println(customer1.getRecord( )==customer2.getRecord( ));
        }
```

8）运行输出结果

运行输出结果如图 13 所示。

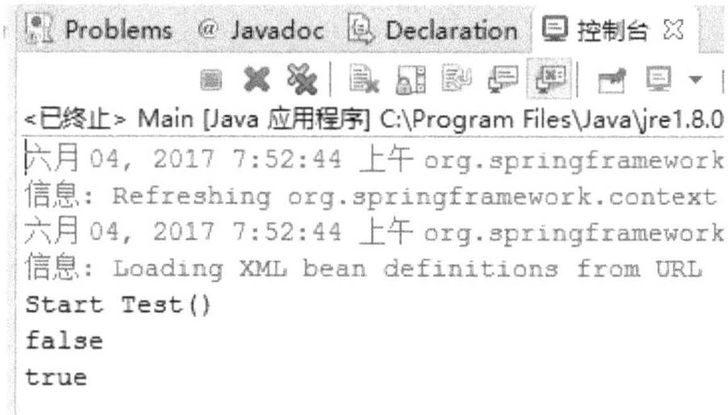

```
Problems  @ Javadoc  Declaration  控制台
<已终止> Main [Java 应用程序] C:\Program Files\Java\jre1.8.0
六月 04, 2017 7:52:44 上午 org.springframework
信息: Refreshing org.springframework.context
六月 04, 2017 7:52:44 上午 org.springframework
信息: Loading XML bean definitions from URL
Start Test()
false
true
```

图 13　运行输出结果

> **专家讲解**
>
> 通过运行结果可知，第一行输出为 false，但是第二行输出为 true。说明 Spring 在底层将类的 record 属性注入两个 Customer 对象。

9）修改为 ref="record2"

将 Customer1 中的 ref="record" 修改为 ref="record2"，再运行单元测试方法，如图 14 所示。

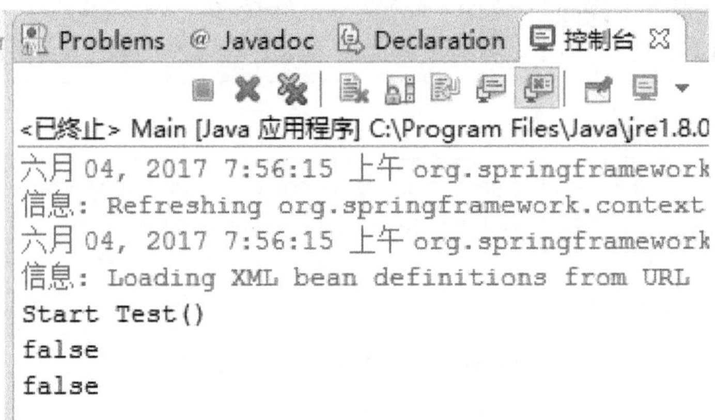

图 14　运行单元测试方法结果

专家讲解

通过运行结果可知，第一行输出为 false，第二行输出也是 false。说明这一次两个对象都是每次新生成一个对象。

可以得出如下结论：Spring 默认将所有的 bean 设置为单例模式，只有当我们特别声明 bean 的模式时，Spring 才会修改对应的 bean 设置，即在存在嵌套关系的 bean 中，即使外层声明为 prototype，其内部在未声明的情况下，仍然是单例的。

2.6　Bean 之间的关系

在前面的内容中，我们介绍了 Spring 的 bean 的各种注入及配置方式。下面介绍 Spring 中 bean 之间的关系，它们之间的关系包括继承、依赖、引用。

2.6.1　继承关系

1）修改 beans.xml 配置文件

修改 beans.xml 配置文件为以下代码。

```
<bean id="AbstractRecord" class="com.iss.demo2.Record" abstract="true">
        <property name="company" value="ABCD"></property>
</bean>
<bean id="Record" parent="AbstractRecord">
        <property name="position" value="Engineer"></property>
        <property name="address" value="Beijing"></property>
</bean>
```

2）修改单元测试方法

修改单元测试方法为以下代码。

```
@org.junit.Test
    public void test1( ) {
        System.out.println("Start Test( )");
        Record customer1 = (Record)ac.getBean("Record");
        System.out.println(customer1);
    }
```

专家讲解

Spring 的继承关系与 Java 的继承关系思想一致，但使用的配置关键字为 abstract。

2.6.2 依赖关系

1）修改 beans.xml

修改 beans.xml 为以下代码。

```
    <bean id="AbstractRecord" class="com.iss.demo2.Record" abstract="true">
        <property name="company" value="ABCD"></property>
    </bean>
    <bean id="Record" parent="AbstractRecord">
        <property name="position" value="Engineer"></property>
        <property name="address" value="Beijing"></property>
    </bean>
    <bean id="Customer1" class="com.iss.demo2.Customer">
        <property name="name" value="Tom"></property>
        <property name="sex" value="male"></property>
        <property name="age" value="22"></property>
    </bean>
</beans>
```

2）修改 Customer.java 和 Record.java 文件

在 Customer.java 和 Record.java 的构造函数中分别加入以下代码。

```
Customer.java 构造函数增加 System.out.println("------ 创建人员 -------");
Record.java 构造函数增加 System.out.println("-------- 创建记录 --------");
```

3）运行输出结果

运行输出结果如图 15 所示。

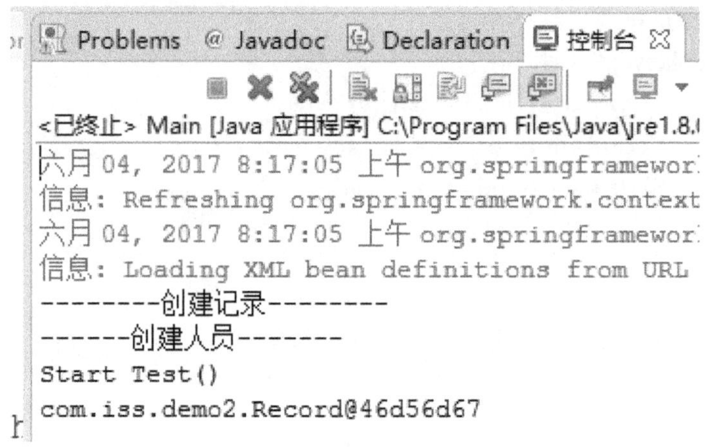

图 15　运行输出结果

专家解释

通过上文可知，创建记录的动作先于创建人员，这是不符合业务逻辑的。因此需要利用 bean 之间的依赖关系来处理此需求。

4）改进方法，修改 id 为 Record 的 bean

修改 id 为 Record 的 bean 的代码如下。

```
<bean id="AbstractRecord" class="com.iss.demo2.Record" abstract="true">
  <property name="company" value="ABCD"></property>
</bean>
<bean id="Record" parent="AbstractRecord" depends-on="Customer1">
  <property name="position" value="Engineer"></property>
  <property name="address" value="Beijing"></property>
</bean>
<bean id="Customer1" class="com.iss.demo2.Customer">
  <property name="name" value="Tom"></property>
  <property name="sex" value="male"></property>
  <property name="age" value="22"></property>
</bean>
```

5）再次运行输出结果

再次运行的输出结果如图 16 所示。

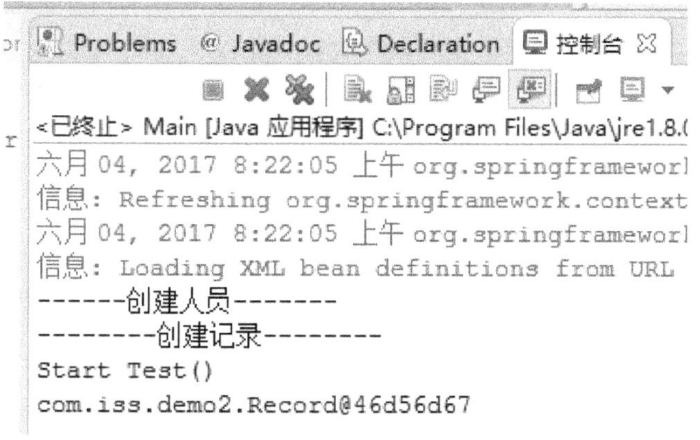

图 16　再次运行输出结果

专家讲解

注意，依赖关系可以多级，多个存在，即如下面的形式。

（1）A → B → C → D。配置方式为：依赖项 A 仅配置相邻依赖项 B，而不配置 C，D 的。

（2）A,B → C → D。配置方式为：依赖项 A 仅配置相邻依赖项 C，依赖项 B 仅配置相邻依赖项 C，而不配置 D 的。

（3）A → B,C → D。配置方式为：依赖项 A 配置相邻依赖项 B 和 C，即使用逗号链接，如 depends-on="idA，idB"。

2.6.3　引用关系

引用关系的使用方法就是我们在前文中多次演示的 ref="id" 的用法，这里不再赘述。

2.7　Bean 的作用域

Spring 中 bean 常用以下两种作用域。

（1）singleton：在每个 Spring IoC 容器中只有一个对象实例，所有引用者共享，Spring 负责 bean 的生命周期。

（2）prototype：每次引用 Spring 都新创建一个 bean，创建后调用者负责销毁对象、回收资源。

复制 Spring_Part2 工程，重命名为 Spring_Part3 工程。修改工程结构如图 17 所示。

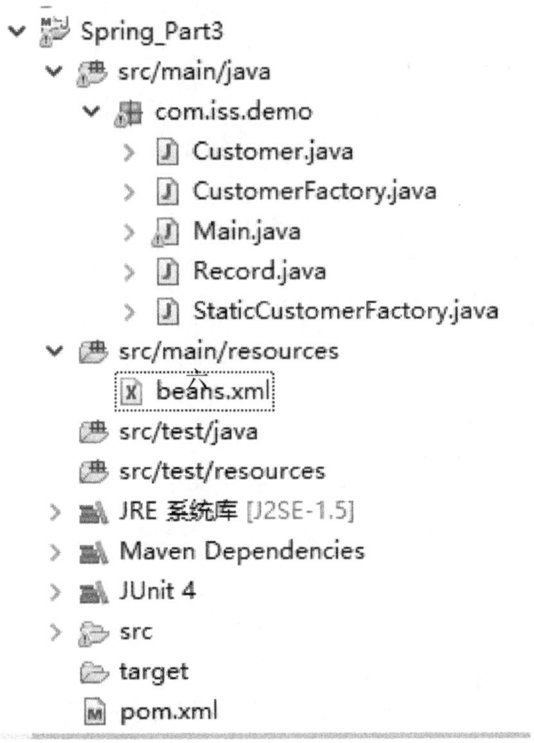

```
Spring_Part3
  src/main/java
    com.iss.demo
      Customer.java
      CustomerFactory.java
      Main.java
      Record.java
      StaticCustomerFactory.java
  src/main/resources
    beans.xml
  src/test/java
  src/test/resources
  JRE 系统库 [J2SE-1.5]
  Maven Dependencies
  JUnit 4
  src
  target
  pom.xml
```

图 17　工程结果图

2.7.1　默认的 singleton 作用域

> **专家提示**
>
> 在每个 Spring IoC 容器，中一个 bean 定义只对应一个对象实例。这个对象实例将会被保存到缓存中，后续对该 bean 的引用和请求都返回该对象实例。

1）修改 beans.xml 文件

```
<bean id="Customer1" class="com.iss.demo.Customer">
        <property name="name" value="Tom"></property>
        <property name="sex" value="male"></property>
        <property name="age" value="22"></property>
</bean>
```

2）修改单元测试方法

```
package com.iss.demo;
public class Main {
private ApplicationContext ac;

```

```
@Before
public void setUp( ) throws Exception {
// 首先读取配置文件,配置文件中的 bean 将会保存到 ApplicationContext 的实例中
ac = new ClassPathXmlApplicationContext("beans.xml");
System.out.println("......junit 单元测试是方法之前执行 ......");
}

 @After
 public void tearDown( ) throws Exception {
 System.out.println("......junit 单元测试是方法之后执行 ......");
 }

 @org.junit.Test
 public void test1( ) {
 System.out.println("Start Test1( )");
 Customer customer1 = (Customer) ac.getBean("Customer1");
 System.out.println(customer1.hashCode( ));

 Customer customer2 = (Customer) ac.getBean("Customer1");
 System.out.println(customer2.hashCode( ));
 }
 }
```

3)运行输出结果

运行输出结果如图 18 所示。

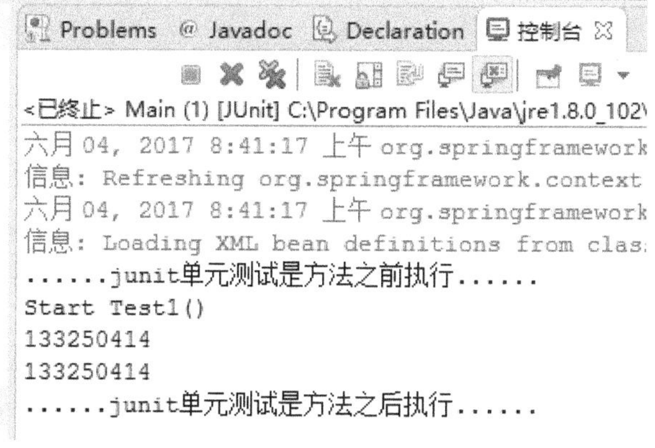

图 18　运行输出结果

专家讲解

这里我们看到 customer1 与 customer2 对应同一个对象，即在默认情况下，Spring 对每一个 bean 只创建一个实例，程序获取的都是这个实例的引用。

2.7.2 prototype 作用域

专家讲解

每次通过容器的 getbean 方法获取 prototype 定义的 bean 时，都将产生一个新的 bean 实例，等价于一个 new 动作。这里需要注意的是：对于配置为 prototype 的 bean，Spring 只负责创建，不负责删除。后续对该对象的生命周期的维护就交给了客户端。

1）修改 beans.xml

```
<bean id="Customer1" class="com.iss.demo.Customer" scope="prototype">
        <property name="name" value="Tom"></property>
        <property name="sex" value="male"></property>
        <property name="age" value="22"></property>
</bean>
```

2）运行输出结果

运行输出结果如图 19 所示。

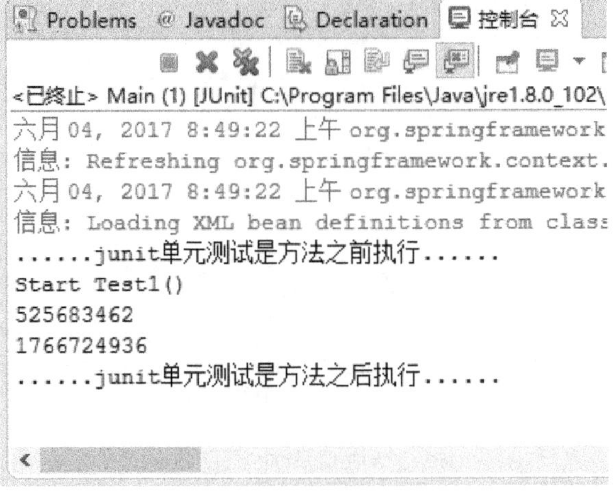

图 19 运行输出结果

专家讲解

这里我们看到 customer1 与 customer2 是不同的对象。如果在 Customer 的构造函数中输出内容的话，那么控制台也可以看到构造函数中的内容被输出了两次，即每次都会获取一个新的 bean 的实例。

2.7.3 request 作用域

专家提示

请在 Web 项目下进行测试，具体细节将在后面的章节中呈现。

在一次 HTTP 请求中，一个 bean 定义对应一个实例；即每次 HTTP 请求将会有各自的 bean 实例，它们依据某个 bean 定义创建而成。该作用域仅在基于 web 的 Spring Application Context 情形下有效。

配置方法为在 web.xml 中增加以下代码。

```
<web-app>
    ...
    <listener>
        <listener-class>
            org.springframework.web.context.request.RequestContextListener
        </listener-class>
    </listener>
    ...
</web-app>
```

beans.xml 的配置代码如下。

```
<bean id="loginAction" class="com.iss.action.LoginAction" scope="request"/>
```

2.7.4 session 作用域

在一个 HTTP session 中，一个 bean 定义对应一个实例。该作用域仅在基于 web 的 Spring Application Context 情形下有效。

beans.xml 的配置代码如下。

```
<bean id="userPreferences" class="com.iss.pojo.UserPreferences" scope="session"/>
```

2.7.5 global 作用域

在一个全局的 HTTP session 中，一个 bean 定义对应一个实例。只是其用于 portlet 环境。

beans.xml 的配置代码如下。

```
<bean id="userPreferences" class="com.iss.pojo.UserPreferences" scope="globalSession"/>
```

2.7.6 application 作用域

同一个 application 共享一个 bean。

beans.xml 的配置代码如下。

```
<bean id="appPreferences" class="com.iss.pojo.AppPreferences" scope="application"/>
```

2.8 配置文件拆分文件策略

2.8.1 为什么需要拆分配置文件

拆分配置文件有以下原因：①项目规模变大，配置文件可读性、可维护性差；②团队开发时，多人修改同一配置文件，易发生冲突。

2.8.2 拆分策略

拆分策略有以下两种。

（1）公用配置 + 每个系统模块有一个单独的配置文件（包含 dao、service、action）。

（2）公用配置 +DAOBean 配置 + 业务逻辑 Bean 配置 +ActionBean 配置。

2.8.3 拆分配置文件的两种方法

拆分配置文件的方法为：配置 ContextLoadListener 的 ContextConfigLocation 属性，配置多个配置文件并用逗号或者通配符“*”隔开。还可以使用 <import resource="x.xml"/> 方式。

beans.xml 代码内容如下。

```
<import resource="beans_sub.xml" />
<!-- <import resource="beans_*.xml" /> 使用通配符方式 ->
```

beans_sub.xml 代码内容如下。

```
<bean id="Customer1" class="com.iss.demo.Customer" scope="prototype">
  <property name="name" value="Tom"></property>
  <property name="sex" value="male"></property>
  <property name="age" value="22"></property>
</bean>
```

小结

控制反转 IoC(Inversion of Control) 是一种设计思想，DI(依赖注入) 是实现 IoC 的一种方法，也有人认为 DI 只是 IoC 的另一种说法。对象的创建由程序自己控制，控制反转后将对象的创建转移给第三方，所谓控制反转就是获得依赖对象的方式给反转了。

IoC 是 Spring 框架的核心内容，它可以有多种实现方式，可以使用 XML 配置，也可以使用注解。新版本的 Spring 也可以零配置实现 IoC。Spring 容器在初始化时先读取配置文件，根据配置文件或元数据创建与组织对象存入容器中，程序使用时再从 IoC 容器中取出需要的对象。

采用 XML 方式配置 bean 时，bean 的定义信息是与实现分离的，而采用注解的方式可以把两者合为一体，bean 的定义信息直接以注解的形式定义在实现类中，从而达到零配置的目的。

经典面试题

（1）根据你的理解，讲一下什么是控制翻转（IoC）?

（2）什么是依赖注入（DI），它有什么作用？

（3）使用 Spring 实现依赖注入的步骤是什么？

（4）项目规模大的时候，byName 方式和 byType 方式哪种更适合？

（5）如何使用 Spring 构造器注入？

（6）Spring 自动装配，采用 byName 和 byType 两种方式的区别是什么？

（7）Spring 的 setter 方式注入与构造器注入实现上的区别是什么？

（8）Spring 采用什么方式简化依赖注入代码和配置文件的编写量？

（9）什么是 Spring bean?

（10）解释 Spring 支持的几种 bean 的作用域。

跟我上机

（1）开发一个打印机，打印机功能的实现依赖于墨盒和纸张（见图 20）。

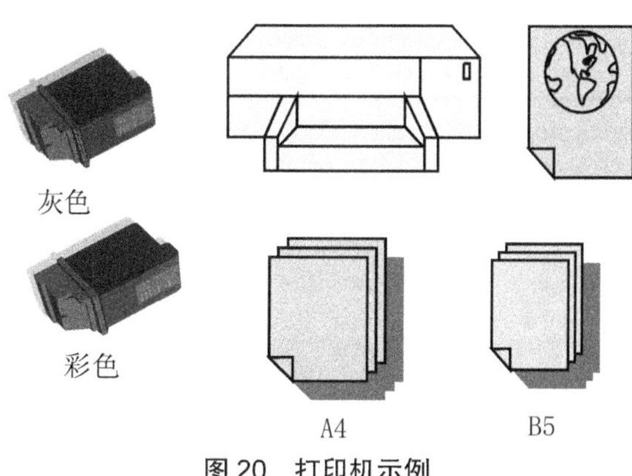

图 20　打印机示例

实现步骤如下：

①定义墨盒和纸张的接口标准；

②使用接口标准开发打印机；

③组装打印机；

④运行打印机。

运行结果如图 21、图 22 所示。

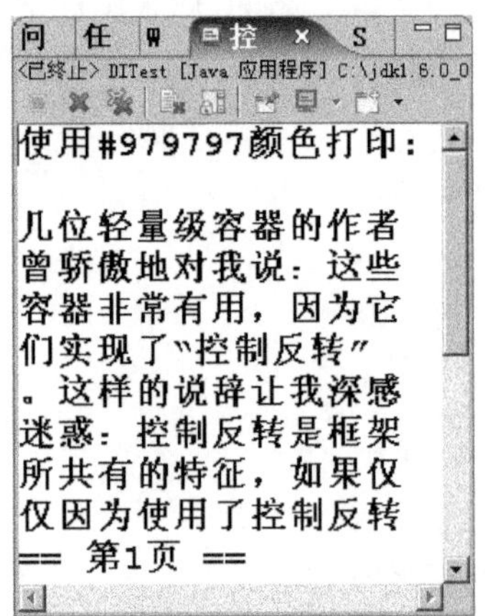

图 21　配置为使用 colorInk、b5Paper 时的运行结果

图 22　配置为使用 greyInk、a4Paper 时的运行结果

代码提示如图 23、图 24 所示。

图 23　代码提示

```
<bean id="colorInk" class="com.iss.springdemo.ink.ColorInk"></bean>
<bean id="grayInk" class="com.iss.springdemo.ink.GrayInk"></bean>
<bean id="a4Paper" class="com.iss.springdemo.paper.TextPaper">
   <property name="charPerLine" value="6"></property>
   <property name="linePerPage" value="5"></property>
</bean>
<bean id="b5Paper" class="com.iss.springdemo.paper.TextPaper">
   <property name="charPerLine" value="10"></property>
   <property name="linePerPage" value="8"></property>
</bean>
<bean id="printer" class="com.iss.springdemo.print.Printer">
   <property name="ink" ref="colorInk"></property>
   <property name="paper" ref="a4Paper"></property>
</bean>
```

图 24　代码提示

（2）通过配置 autowire 为 byName 和 byType 完成电脑开机的案例，在控制台输出信息（图 24）。

```
Console
<terminated> Main (1) [Java Application] C:\Program Files\Java\jre1.8.0_102\bin\javaw.exe (2017年6月4日 下午9:35:46)
*****************************************
    autowired=byName

  我是主机 正在运行！ 我是显示器 正在运行

    autowired=byType

  我是主机 正在运行！ 我是显示器 正在运行

*****************************************
```

图 25　输出信息

（3）拆分配置文件。把某系统中的 Spring 配置文件 applicationContext.xml 拆分成以下四个配置文件：

① action.xml（单独配置 action bean）；

② dao.xml（单独配置 dao bean）；

③ service.xml（单独配置 service bean）；

④ core.xml (配置数据源和 SessionFactory）。

通过一个 applicationContext.xml 来引用上面四个配置文件。

第3章　Spring 注解配置 IoC

本章要点（学会后请在方框中打钩）：

- □ 掌握使用 Spring 注解配置 IoC
- □ 掌握自动装配
- □ 掌握零配置实现 IoC
- □ 掌握 Spring 各种注解的使用

3.1 使用注解配置 IoC

上一章是使用传统的 xml 配置完成 IoC 的，如果内容比较多则需花费很多时间，通过注解可以减轻工作量，但注解后的修改要麻烦一些，耦合度会增加，应该根据需要选择合适的方法。

假设项目中需要完成对图书的数据访问服务，现已定义好 IBookDAO 接口与 BookDAO 实现类。

IBookDAO.java 的代码内容如下。

```
package com.iss.dao;
public interface IBookDAO {
/**
 * 添加图书
 */
public String addBook(String bookname);
}
```

BookDAO.java 的代码内容如下。

```
package com.iss.dao.impl;
import com.iss.dao.IBookDAO;
public class BookDAO implements IBookDAO {
public String addBook(String bookname) {
        return " 添加图书 " + bookname + " 成功！ ";
}
}
```

Maven 项目工程文件结构如图 1 所示。

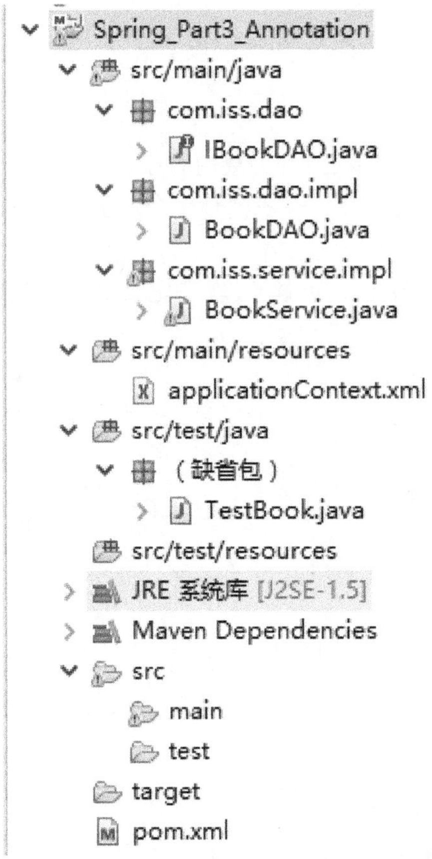

```
Spring_Part3_Annotation
  src/main/java
    com.iss.dao
      IBookDAO.java
    com.iss.dao.impl
      BookDAO.java
    com.iss.service.impl
      BookService.java
  src/main/resources
    applicationContext.xml
  src/test/java
    (缺省包)
      TestBook.java
  src/test/resources
  JRE 系统库 [J2SE-1.5]
  Maven Dependencies
  src
    main
    test
  target
  pom.xml
```

图 1 Maven 项目工程文件结构

3.1.1 修改 BookDAO.java

```
package com.iss.dao.impl;
import org.springframework.stereotype.Component;
import com.iss.dao.IBookDAO;

@Component("bookDAO")
public class BookDAO implements IBookDAO {
public String addBook(String bookname) {
  return " 添加图书 " + bookname + " 成功！ ";
}
}
```

专家讲解

在类上增加了一个注解 Component，在类的开头使用了 @Component 注解，它可以被 Spring 容器识别，启动 Spring 后，会自动把它转成容器管理的 Bean。

除了 @Component 外，Spring 提供了 3 个功能基本和 @Component 等效的注解，分别用于对 DAO、Service 和 Controller 进行注解。

（1）@Repository 用于对 DAO 实现类进行注解。

（2）@Service 用于对业务层注解，但是目前该功能与 @Component 相同。

（3）@Constroller 用于对控制层注解，但是目前该功能与 @Component 相同。

3.1.2 修改 BookService

```
package com.iss.service.impl;

public class BookService {
   IBookDAO bookDAO;
   public BookService( ) {
         // 容器
         ApplicationContext ctx = new
ClassPathXmlApplicationContext("applicationContext.xml");
         // 从容器中获得 id 为 bookdao 的 bean
         bookDAO = (IBookDAO) ctx.getBean("bookDAO");
}

public void storeBook(String bookname) {
         System.out.println(" 图书上货 ");
         String result = bookDAO.addBook(bookname);
         System.out.println(result);
}
}
```

将构造方法中的代码直接写在 storeBook 方法中，以避免循环加载的问题。

3.1.3 修改 applicationContext.xml

```
<?xml version="1.0" encoding="UTF-8"?>
<beans xmlns="http://www.springframework.org/schema/beans"
      xmlns:xsi="http://www.w3.org/2001/XMLSchema-instance"
      xmlns:p="http://www.springframework.org/schema/p"
      xmlns:context="http://www.springframework.org/schema/context"
      xsi:schemaLocation="http://www.springframework.org/schema/beans
            http://www.springframework.org/schema/beans/spring-beans.xsd
```

```
             http://www.springframework.org/schema/context
             http://www.springframework.org/schema/context/spring-context-4.3.xsd">
             <context:component-scan base-package="com.iss.dao.impl">
             </context:component-scan>
</beans>
```

专家讲解

粗体字是新增的 xml 命名空间与模式约束文件位置。增加了注解扫描的范围，指定了一个包，可以通过属性设置更加精确的范围。

3.1.4 编写测试类

```
import org.junit.Test;
import com.iss.service.impl.BookService;
public class TestBook {
   @Test
   public void testBook( ) {
         BookService bookservice = new BookService( );
         bookservice.storeBook("《Spring+Spring MVC+MyBatis 全解 Version》");
   }
}
```

3.1.5 运行输出结果

运行输出结果如图 2 所示。

```
Problems  @ Javadoc  Declaration  控制台
<已终止> TestBook [JUnit] C:\Program Files\Java\jre1.8.0_102\bin\javaw.exe
六月 04, 2017 11:18:35 下午 org.springframework.context.s
信息: Refreshing org.springframework.context.support.Cla
六月 04, 2017 11:18:35 下午 org.springframework.beans.fac
信息: Loading XML bean definitions from class path resou
图书上货
添加图书《Spring+Spring MVC+MyBatis 全解 Version》成功!
```

图 2 运行输出结果

专家讲解

从配置文件中可以看出，我们并没有声明 BookDAOObj 与 BookService 类型的对象，但还是从容器中获得实例并成功运行了，原因是在类的开头使用了 @Component 注解，它可以被 Spring 容器识别，启动 Spring 后，会自动把它转成容器管理的 bean。

3.2 使用注解自动装配

从上一个示例可以看出，有两个位置都使用 ApplicationContext 初始化容器后获得需要的 bean，可以通过自动装配进行简化。

3.2.1 修改 BookDAO.java

```
@Repository
public class BookDAO implements IBookDAO {
    public String addBook(String bookname) {
        return " 添加图书 "+bookname+" 成功！ ";
    }
}
```

专家讲解

把注解修改成了 Repository，比 Component 更贴切一些，但非必要。

3.2.2 修改 BookService.java

```
@Service
public class BookService {
    @Autowired
    IBookDAO bookDAO;
    public void storeBook(String bookname){
        System.out.println(" 图书上货 ");
        String result=bookDAO.addBook(bookname);
        System.out.println(result);
    }
}
```

专家讲解

将类 BookService 上的注解替换成了 Service；在 bookDAO 成员变量上增加了一个注解 @Autowired，该注解的作用是：可以对成员变量、方法和构造函数进行注解，来完成自动装配的工作，通俗来说就是会根据类型从容器中自动查找到一个 bean 给 bookDAO 字段。@Autowired 是根据类型进行自动装配的，如果需要按名称进行装配，则需要配合 @Qualifier。另外还可以使用其他注解，@ Resource 等同于 @Qualifier，@Inject 等同于 @ Autowired。

专家讲解

@Service 用于注解业务层组件（我们通常定义的 Service 层就用这个）。

@Controller 用于注解控制层组件（如 struts 中的 action)。

@Repository 用于注解数据访问组件，即 DAO 组件。

@Component 泛指组件，当组件不好归类的时候，可以使用这个注解进行注解。

专家讲解

（1）@Resource 默认是按照名称装配注入的，只有在找不到与名称匹配的 bean 时才会按照类型来装配注入。

（2）@Autowired 默认是按照类型装配注入的，如果想按照名称来装配注入，则需要结合 @Qualifier 一起使用。

（3）@Resource 注解是由 Java EE 提供的，而 @Autowired 是由 Spring 提供的，为减少系统对 Spring 的依赖建议使用 @Resource 的方式；如果 Maven 项目是 1.5 的 JRE 则需换成更高版本的。

（4）@Resource 和 @Autowired 都可以书写注解在字段或者该字段的 setter 方法之上。

（5）@Autowired 可以对成员变量、方法以及构造函数进行注解，而 @Qualifier 的注解对象是成员变量、方法入参、构造函数入参。

（6）@Qualifier("XXX") 中的 XXX 是 bean 的名称，所以 @Autowired 和 @Qualifier 结合使用时，自动注入的策略就从 byType 转变成 byName 了。

（7）@Autowired 注解进行自动注入时，Spring 容器中匹配的候选 bean 数目必须有且仅有一个，通过属性 Required 可以设置为非必要。

（8）@Resource 装配顺序。

如果同时指定了 Name 和 Type，则将从 Spring 上下文中找到唯一匹配的 bean 进行装配，找不到则抛出异常。

如果指定了 Name，则从上下文中查找名称（id）匹配的 bean 进行装配，找不到则抛出异常。

如果指定了 Type，则从上下文中找到类型匹配的唯一 bean 进行装配，找不到或者找到多个，都会抛出异常。

如果既没有指定 Name，又没有指定 Type，则自动按照 byName 方式进行装配；如果没有匹配，则回退为一个原始类型进行匹配，如果匹配则自动装配。

专家举例

```
@Service
public class BookService {
    public IBookDAO getDaoofbook( ) {
        return daoofbook;
    }
```

```
    /*
    @Autowired
    @Qualifier("bookdao02")
    public void setDaoofbook(IBookDAO daoofbook) {
            this.daoofbook = daoofbook;
    }*/

    @Resource(name="bookdao02")
    public void setDaoofbook(IBookDAO daoofbook) {
           this.daoofbook = daoofbook;
    }

    /*
  @Autowired
  @Qualifier("bookdao02")
   */
   IBookDAO daoofbook;

   /*
   public BookService(@Qualifier("bookdao02") IBookDAO daoofbook) {
         this.daoofbook=daoofbook;
   }*/

   public void storeBook(String bookname){
         System.out.println(" 图书上货 ");
         String result=daoofbook.addBook(bookname);
         System.out.println(result);
   }
}
```

3.2.3 编写测试类

```
public class TestBook {
@Test
public void testBook( ) {
      ApplicationContext ctx = new
  ClassPathXmlApplicationContext("applicationContext.xml");
```

```
        BookService bookservice = ctx.getBean(BookService.class);
        bookservice.storeBook("《Spring+Spring MVC+MyBatis 全解 Version》");
    }
}
```

3.2.4 运行输出结果

运行输出结果如图 3 所示。

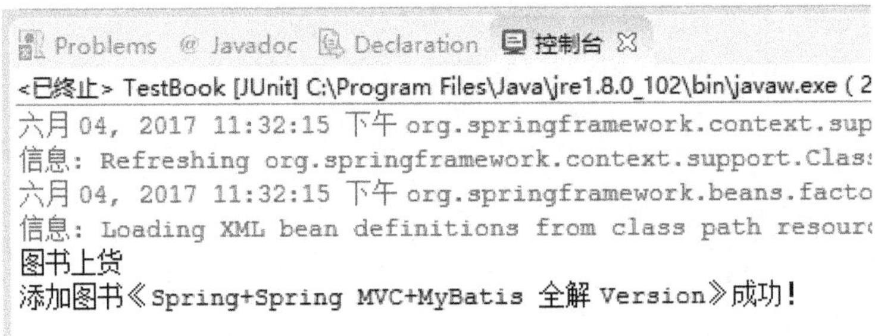

```
Problems  @ Javadoc  Declaration  控制台 ☒
<已终止> TestBook [JUnit] C:\Program Files\Java\jre1.8.0_102\bin\javaw.exe ( 2
六月 04, 2017 11:32:15 下午 org.springframework.context.sup
信息: Refreshing org.springframework.context.support.Clas:
六月 04, 2017 11:32:15 下午 org.springframework.beans.facto
信息: Loading XML bean definitions from class path resour
图书上货
添加图书《Spring+Spring MVC+MyBatis 全解 Version》成功!
```

图 3　运行输出结果

3.3 零配置实现 IoC

所谓的零配置就是不再使用 xml 文件来初始化容器，而是使用一个类型来替代。

3.3.1 编写 IBookDAO.java

```
public interface IBookDAO {
        public String addBook(String bookname);
}
```

3.3.2 编写实现类 BookDAO.java

```
@Repository
public class BookDAO implements IBookDAO {
        public String addBook(String bookname) {
                return " 添加图书 "+bookname+" 成功！ ";
        }
}
```

在 BookDAO 类上注解了 @Repository，当初始化时该类将被容器管理，会生成一个 bean，可以通过构造方法测试。

3.3.3 编写业务层 BookService.java

```
@Service
public class BookService {
    @Resource
    IBookDAO bookDAO;
    public void storeBook(String bookname){
        System.out.println(" 图书上货 ");
        String result=bookDAO.addBook(bookname);
        System.out.println(result);
    }
}
```

专家讲解

因为注解了 @Service 所以类 BookService 将对容器进行管理，初始化时会生成一个单例的 bean，类型为 BookService。在字段 bookDAO 上注解了 @Resource，用于自动装配，Resource 默认是按照名称来装配注入的，只有当找不到与名称匹配的 bean 时才会按照类型来装配注入。

3.3.4 编写 ApplicationCfg 类

```
@Configuration
@ComponentScan(basePackages="com.iss.dao.impl,com.iss.service.impl")
public class ApplicationCfg {
    @Bean
    public User getUser( ){
        return new User(" 成功 ");
    }
}
```

专家讲解

@Configuration 相当于配置文件中的 <beans/>。

@ComponentScan 相当于配置文件中的 context:component-scan，属性也一样设置。

@Bean 相当于 <bean/>，只能注解在方法和注解上，一般在方法上使用。

@Target({ElementType.METHOD, ElementType.ANNOTATION_TYPE})，方法名相当于 id。

3.3.5 编写 User.java

```
@Component("user1")
public class User {
        public User( ) {
              System.out.println(" 创建 User 对象 ");
        }
        public User(String msg) {
              System.out.println(" 创建 User 对象 "+msg);
        }
        public void show( ){
              System.out.println(" 一个学生对象！ ");
        }
}
```

3.3.6 编写测试类

```
// 初始化容器的代码与以前有一些不一样，具体如下：
public class Test {
        @org.junit.Test
        public void testStoreBook( )
        {
  // 容器，注解配置应用程序容器，Spring 通过反射 ApplicationCfg.class 初始化容器
            ApplicationContext ctx=new AnnotationConfigApplicationContext(ApplicationCfg.class);
            BookService bookservice=ctx.getBean(BookService.class);
            bookservice.storeBook("《Spring+Spring MVC+MyBatis 全解 Version》");
              User user1=ctx.getBean("user1",User.class);
              user1.show( );
              User getUser=ctx.getBean("getUser",User.class);
              getUser.show( );
        }
}
```

3.3.7 运行输出结果

运行输出结果如图 4 所示。

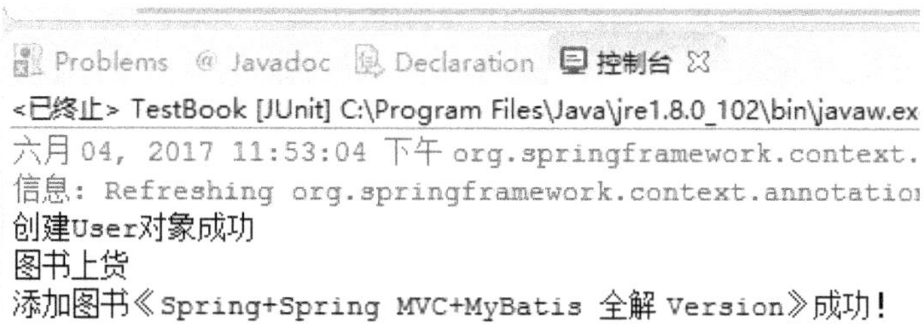

```
Problems @ Javadoc Declaration 控制台
<已终止> TestBook [JUnit] C:\Program Files\Java\jre1.8.0_102\bin\javaw.ex
六月 04, 2017 11:53:04 下午 org.springframework.context.
信息: Refreshing org.springframework.context.annotatio
创建User对象成功
图书上货
添加图书《Spring+Spring MVC+MyBatis 全解 Version》成功!
```

图 4 运行输出结果

小结

使用零配置和注解虽然方便,不需要编写麻烦的 xml 文件,但这样做并非是为了取代 xml,应该根据实际需要进行选择,或二者结合使用,毕竟使用一个类作为容器的配置信息是硬编码的,在发布后不易修改。

经典面试题

(1)为什么要使用 Spring 注解?

(2)Spring 注解需要引用哪些包?

(3)列举 Spring 中常用的注解。

(4)Spring 注解方式和配置方式的区别是什么?能否用注解方式和配置方式的代码分别举例?

(5)如何解释 @Service 注解?

(6)如何解释 @Controller 注解?

(7)如何解释 @Repository 注解?

(8)如何解释 @Component 注解?

(9)如何解释 @Resource 注解?

(10)如何解释 @Autowire 注解?

跟我上机

(1)建立 Maven Web 项目,制作一个简单的公司人员权限管理功能。

按表 1 要求创建数据库和表。

表 1　表数据

数据库名	PeopleManage		表名	Users
字段显示	字段名	数据类型	字段大小	备注和说明
编号	ID	int		标识列，主键
姓名	Name	Varchar	50	员工姓名
登录名	LoginId	Varchar	50	登录名
停用状态	DeleteState	int		0 为停用，1 为启用
角色编号	UserRoleId	int		1 为管理员，2 为普通用户

显示结果功能如图 5 所示。

用户表

姓名	登录名	是否停用	角色	操作
CEO	ceo	停用	普通用户	删除 启用
CFO	cfo	启用	管理员	删除 停用
CTO	ufo	停用	管理员	删除 启用
UFO	ufo	启用	普通用户	删除 停用
张三	rr	启用	普通用户	删除 停用

10　第 1 共1页　显示1到5,共5记录

图 5　显示结果

第 4 章　Spring AOP(面向切面编程)

本章要点(学会后请在方框中打钩):

- □ 了解什么是 AOP(面向切面编程)
- □ 掌握基于 xml 配置的 Spring AOP
- □ 掌握使用注解配置 AOP
- □ 理解 AspectJ 切点函数
- □ 理解 AspectJ 通知注解
- □ 掌握零配置实现 Spring IoC 与 AOP
- □ 掌握 Spring 提供的各种增强处理类型

4.1 了解 AOP

在软件业，AOP 为 Aspect Oriented Programming 的缩写，意为：面向切面编程，是通过预编译方式和运行时的动态代理实现程序功能统一维护的一种技术。

AOP 是 OOP 的延续，是软件开发中的一个热点，也是 Spring 框架中的一个重要内容，是函数式编程的一种衍生范型。

利用 AOP 可以对业务逻辑的各个部分进行隔离，从而使得业务逻辑各部分之间的耦合度降低，提高程序的可重用性，同时提高了开发的效率。

专家讲解

AOP 就是我们把一个方法看做一个切面，在这个切面的前后或者周围，都可以设置其他的处理方法，进行一些特殊的处理。

比如“吃蛋糕”方法，在使用这个方法前需要“拆包装”，使用这个方法后需要“打扫卫生”，都可以通过这种编程方式来实现。

4.1.1 实现主要功能

使用 AOP 能够实现日志记录、性能统计、安全控制、事务处理、异常处理等功能。

4.1.2 AOP 的目标

使用 AOP 让我们可以“专心做事”。

4.1.3 AOP 原理

AOP 的原理是将复杂的需求分解成不同切面，将散布在系统中的公共功能集中解决。采用代理机制组装起来运行，在不改变原程序的基础上对代码段进行增强处理或增加新的功能等。

4.1.4 AOP 的业务逻辑运行图

图 1 所示示例中，实际运行的流程被划分为不同的执行模块，各个模块之间界限分明，没有出现业务逻辑交织在一起的情况，并且各个模块之间原有的设计方式并没有改变，因此也不会干扰到原有程序的执行过程。

AOP 可以分为静态织入与动态织入，静态织入即在编译前将需织入的内容写入目标模块中，但是成本非常高，动态织入则不需要改变目标模块。Spring 框架实现了 AOP，使用注解配置完成 AOP 比使用 XML 配置要更加方便与直观。

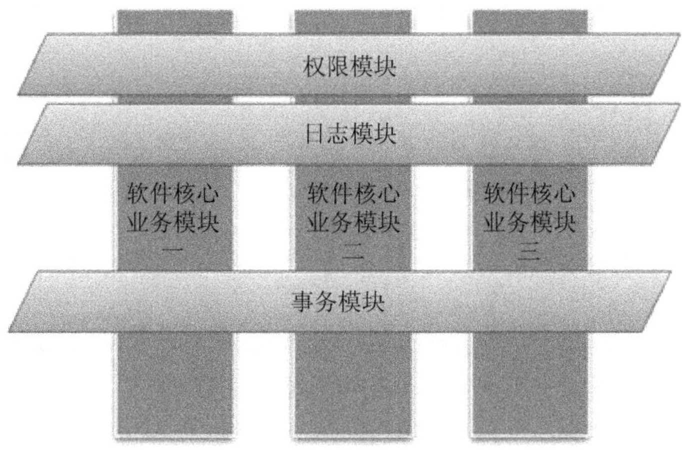

图 1　AOP 的业务逻辑运行图

4.1.5　综合实例演示

如果要实现注册一个用户的功能，我们需要记录这个环节的日志，那么日志处理的代码将会被加入注册的方法中，如下所示。

```
@Override
    public void registerUser(MgrUser user) {
        //logger.info(" 开始：保存用户信息方法 ")
        // 获取用户信息
        // 存储用户信息
        // 返回存储结果
        //logger.info(" 结束：保存用户信息方法 ")
}
```

专家讲解

上面的例子中，日志处理的代码已经进入业务流程的处理代码，如果换成其他的功能，如用户身份验证、权限验证、频率统计等，各种各样的代码都会交织在一起。这样的代码无论是从维护的角度还是程序架构的角度都是不合适的。因此，这里我们就需要 AOP 的概念来帮助我们将各个功能的代码拆分开来。各个模块的代码独立，并且能够按照一定的顺序执行。

专家讲解

“面向切面”的概念，可以类比 Java 中的“面向对象”，类比数学上的两个形状不同的图形“相切”来理解。即我们设计的两段代码是职责单一的“对象”，业务功能相互分离。但在执行顺序上，两者相邻，相互之间存在着一个“切面”。Spring 将系统中的应用分为两个大的部分：核心业务逻辑和横向的通用逻辑。这些通用的逻辑就是我们上文中提到的持久化管理、事务管理、安全管理、日志管理等。

下面通过实例来说明。

1）新建 Maven 工程

新建 Maven 工程，完成功能后工程结构如图 2 所示。

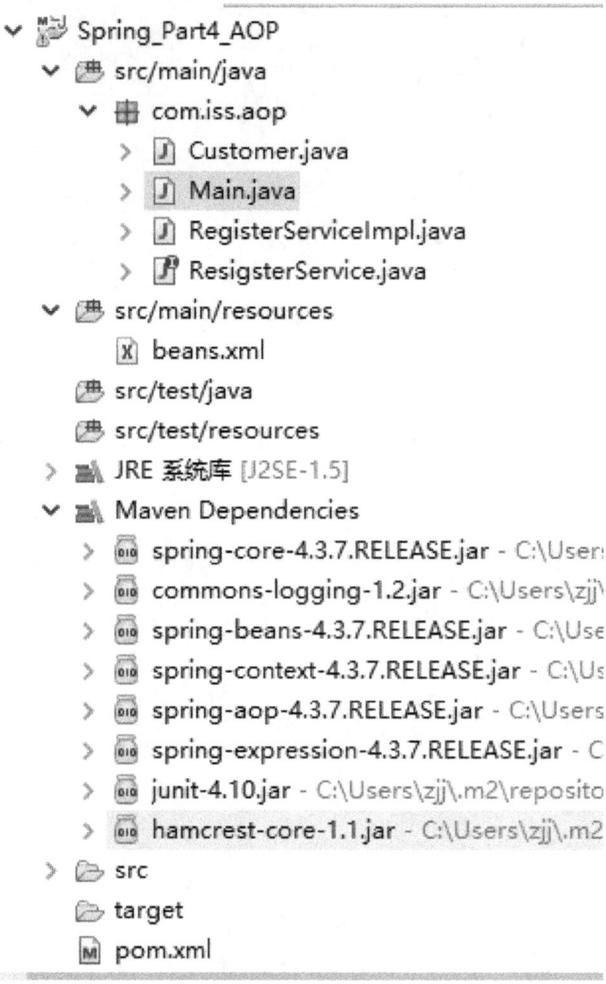

```
Spring_Part4_AOP
  src/main/java
    com.iss.aop
      Customer.java
      Main.java
      RegisterServiceImpl.java
      ResigsterService.java
  src/main/resources
    beans.xml
  src/test/java
  src/test/resources
  JRE 系统库 [J2SE-1.5]
  Maven Dependencies
    spring-core-4.3.7.RELEASE.jar - C:\User:
    commons-logging-1.2.jar - C:\Users\zjj\
    spring-beans-4.3.7.RELEASE.jar - C:\Use
    spring-context-4.3.7.RELEASE.jar - C:\Us
    spring-aop-4.3.7.RELEASE.jar - C:\Users
    spring-expression-4.3.7.RELEASE.jar - C
    junit-4.10.jar - C:\Users\zjj\.m2\reposito
    hamcrest-core-1.1.jar - C:\Users\zjj\.m2
  src
  target
  pom.xml
```

图 2　工程结构图

2）定义 Customer.java

```
public class Customer {
private String name;
private String sex;
private int age;
// 省略 getter 和 setter
@Override
public String toString( ) {
```

```
        return "Customer [name=" + name + ", sex=" + sex + ", age=" + age + "]";
    }
}
```

3）创建 ResigsterService.java

```
package com.iss.aop;
public interface ResigsterService {
        public void resigsterCustomer(Customer c);
}
```

4）创建 RegisterServiceImpl.java

```
package com.iss.aop;

public class RegisterServiceImpl implements ResigsterService {
public void resigsterCustomer(Customer c) {
         System.out.println(" 调用 dao 层，存数客户 "+c.getName( ));
}
}
```

5）修改 beans.xml

```
<?xml version="1.0" encoding="UTF-8"?>
<beans xmlns="http://www.springframework.org/schema/beans"
xmlns:xsi="http://www.w3.org/2001/XMLSchema-instance"
xsi:schemaLocation="http://www.springframework.org/schema/beans
            http://www.springframework.org/schema/beans/spring-beans.xsd">
<bean id="Register" class="com.iss.aop.RegisterServiceImpl">
</bean>
</beans>
```

6）修改测试方法

```
package com.iss.aop;
import org.junit.After;
import org.junit.Before;
// 导入类略
public class Main {
private ApplicationContext ac;
@Before
```

```
public void setUp( ) throws Exception {
    // 首先读取配置文件，配置文件中的 bean 将会保存到 Application Context 的实例中
    ac = new ClassPathXmlApplicationContext("beans.xml");
    System.out.println("......junit 单元测试是方法之前执行 ......");
}
@After
public void tearDown( ) throws Exception {
    System.out.println("......junit 单元测试是方法之后执行 ......");
}
@org.junit.Test
public void test1( ) {
    ResigsterService resigster = (ResigsterService) ac.getBean("Register");
    resigster.resigsterCustomer(new Customer("Tom", "male", 22));
}
}
```

7）运行输出结果

运行输出结果如图 3 所示。

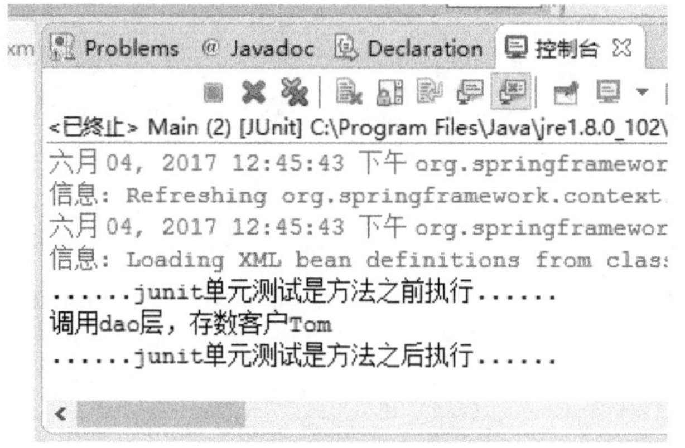

图 3　运行输出结果

专家讲解

上面的结果显示了使用 Junit 单元测试的显示结果，注意单元测试 @Before 和 @After 的用法。以后会与 Spring 中的 @Before 和 @After 进行对比。

接下来看一下当需要日志中的场景，进行用户注册时，需要记录的日志信息。

8）修改 RegisterServiceImpl.java

```
public void resigsterCustomer(Customer c) {
        System.out.println("----------- 调用 Log, 保存日志 ------------");
        System.out.println(" 开始注册用户："+c.getName( ));
        System.out.println("----------- 调用 Log, 保存日志 ------------");
        System.out.println(" 调用 dao 层，存数客户 "+c.getName( ));
        System.out.println("----------- 调用 Log, 保存日志 ------------");
        System.out.println(" 结束注册用户："+c.getName( ));
        System.out.println("----------- 调用 Log, 保存日志 ------------");
}
```

9）运行输出结果

运行输出结果如图 4 所示。

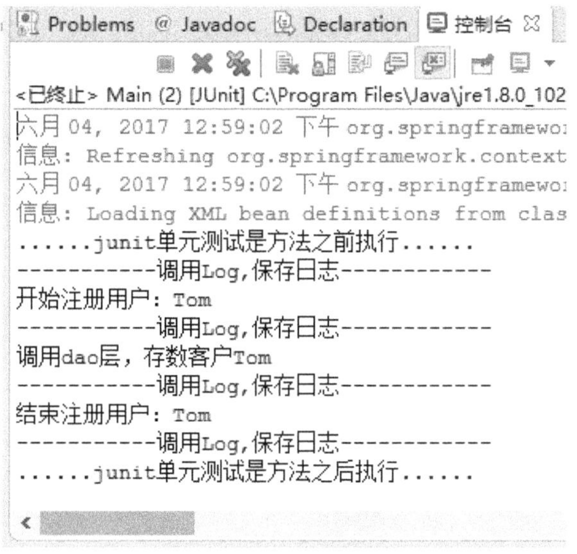

图 4　运行输出结果

专家讲解

上面的示例中日志的代码已经进入到业务逻辑当中，代码的耦合度非常高。这是一种非常不合理的做法。我们要把上面方法中的多个任务分开，分解成多个方法。如何分解，先看一下 4.2 的解释。

4.2　注解分类和注解 AOP

（1）切面 (aspect)：用来切插业务方法的类。如上文中将日志代码剥离出来成为一个单独的类，在需要使用的时候，自动调用这个类。

（2）连接点 (joinpoint)：是切面类和业务类的连接点，其实就是封装了业务方法的一些基本属性，作为通知的参数来解析。

（3）通知 (advice)：在切面类中，声明对业务方法做额外处理的方法。如，通知切面类的运行结果等。

（4）切入点 (pointcut)：业务类中指定的方法，作为切面切入的点。其实就是指定某个方法作为切面切的地方。如上文提到的 resigsterCustomer 方法。

4.2.1 通知的分类

（1）前置通知 (before advice)：在切入点之前执行。

（2）后置通知 (after returning advice)：在切入点执行完成后，执行通知。

（3）环绕通知 (around advice)：包围切入点，调用方法前后完成自定义行为。

（4）异常通知 (after throwing advice)：在切入点抛出异常后，执行通知。

4.2.2 注解配置 AOP

本节我们使用注解配置 AOP 切面，完成对 Business 类的 delete 方法进行操作时，做前置增强，后置增强，异常增强处理和环绕增强处理，建立 Spring_AOP_3，工程基本结构，如图 5 所示。

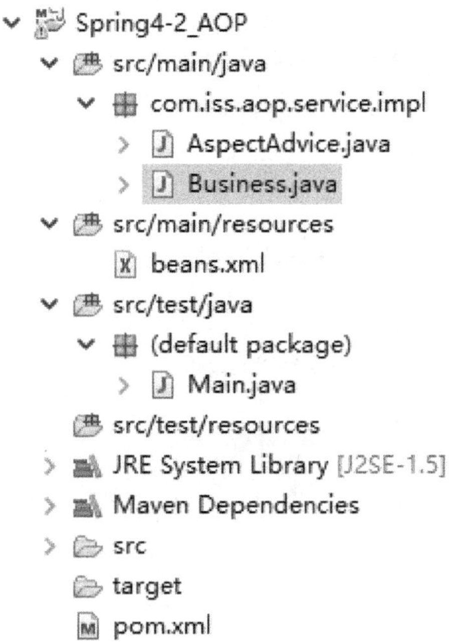

```
Spring4-2_AOP
  src/main/java
    com.iss.aop.service.impl
      AspectAdvice.java
      Business.java
  src/main/resources
    beans.xml
  src/test/java
    (default package)
      Main.java
  src/test/resources
  JRE System Library [J2SE-1.5]
  Maven Dependencies
  src
  target
  pom.xml
```

图 5 工程结构图

1）创建 Business.java 文件

```java
package com.iss.aop.service.impl;
@Component // 必须要有
public class Business {
        /**
         * 切入点
         */
        public String delete(String obj) {
            System.out.println("==========delete===========");
            return obj;
        }

        public String add(String obj) {
            System.out.println("================add==============");
            return obj ;
        }

        public String modify(String obj) {
            System.out.println("==========modify============");
            return obj;
        }
}
```

2）创建 beans.xml 文件

```xml
<context:component-scan base-package="com.iss.aop" />
<!-- 打开 aop 注解 -->
<aop:aspectj-autoproxy />
```

3）创建 Main.java 文件

```java
package Main;
public class Main {
        @SuppressWarnings("resource")
        public static void main(String[] args) {
        ApplicationContext context = new ClassPathXmlApplicationContext("beans.xml");
              Business business = (Business) context.getBean("business");
              business.delete("JACK");
```

```
    }
}
```

4）创建 AspectAdvice.java 文件

```
package com.iss.aop.service.impl;
@Component
@Aspect
public class AspectAdvice {
    /**
     * 指定切入点匹配表达式，注意它是以方法的形式进行声明的。
     */
    @Pointcut("execution(* com.iss.aop.service.impl.*.*(..))")
    public void anyMethod( ) {
    }

    /**
     * 前置通知
     */
  @Before(value = "execution(* com.iss.aop.service.impl.*.*(..))")
  public void doBefore(JoinPoint jp) {
     System.out.println("== 进入 before advice==");
     System.out.print(" 准备在 " + jp.getTarget( ).getClass( ) + " 对象上用 ");
     System.out.print(jp.getSignature( ).getName( ) + " 方法进行对 '");
     System.out.print(jp.getArgs( )[0] + "' 进行删除！ \n\n");
     System.out.println("== 退出 before advice==");
  }

  /**
   * 后置通知
   */
  @AfterReturning(value = "anyMethod( )", returning = "result")
  public void doAfter(JoinPoint jp, String result) {
        System.out.println("== 进入 after advice==");
        System.out.println(" 切入点方法执行完了 ");
        System.out.print(jp.getArgs( )[0] + " 在 ");
        System.out.print(jp.getTarget( ).getClass( ) + " 对象上被 ");
        System.out.print(jp.getSignature( ).getName( ) + " 方法删除了 ");
```

```
        System.out.print(" 只留下：" + result + "\n");
        System.out.println("== 退出 after advice==");
    }

    /**
     * 环绕通知
     */
    @Around(value = "execution(* com.iss.aop.service.impl.*.*(..))")
    public void doAround(ProceedingJoinPoint pjp) throws Throwable {
        System.out.println("== 进入 around 环绕方法！ ==\n");
        // 调用目标方法之前执行的动作
        System.out.println(" 调用方法之前：执行 ");
        // 调用方法的参数
        Object[] args = pjp.getArgs( );
        // 调用的方法名
        String method = pjp.getSignature( ).getName( );
        // 获取目标对象
        Object target = pjp.getTarget( );
        // 执行完方法的返回值：调用 proceed( ) 方法，就会触发切入点方法执行
        Object result = pjp.proceed( );
        System.out.println(" 输出：" + args[0] + ";" + method + ";" + target + ";" + result );
        System.out.println("== 退出 around 环绕方法！ ==");
    }

    /**
     * 异常通知
     */
    @AfterThrowing(value = "execution(* com.iss.aop.service.impl.*.*(..))", throwing = "e")
    public void doThrow(JoinPoint jp, Throwable e) {
        System.out.println(" 删除出错啦 ");
    }
}
```

5）测试运行输出结果

测试运行输出结果如图 6 所示。

图 6 测试运行输出结果

4.3 Spring AOP 的 execution 表达式

切面表达式常见的书写方法如下。

```
execution(* com.iss.service.impl..*.*(..))// 写在方法上
```

符号相关释意如表 1 所示。

表 1 符号相关释意

符号	含义
execution()	表达式的主体
第一个“*”符号	表示返回值的类型任意
com.iss.service.impl	AOP 所切的服务的包名，即，我们的业务部分
包名后面的“..”	表示当前包及子包
第二个“*”	表示类名，* 即所有类。此处可以自定义，下文有举例
.*(..)	表示任何方法名，括号表示参数，两个点表示任何参数类型

4.3.1 execution 表达式详解

execution 通用语法格式如下。

```
execution(< 修饰符模式 >?< 返回类型模式 >< 方法名模式 >(< 参数模式 >)< 异常模式 >?)
```

除了返回类型模式、方法名模式和参数模式外，其他项都是可选的。

下面给出一些通用切入点表达式的例子。

（1）任意公共方法的执行。

```
execution(public * *(..))
```

（2）任何一个名字以"set"开始的方法的执行。

```
execution(* set*(..))
```

（3）AccountService 接口定义的任意方法的执行。

```
execution(* com.iss.service.AccountService.*(..))
```

（4）在 service 包中定义的任意方法的执行。

```
execution(* com.iss.service.*.*(..))
```

（5）在 service 包或其子包中定义的任意方法的执行。

```
execution(* com.iss.service..*.*(..))
```

（6）在 service 包中的任意连接点（在 Spring AOP 中只是方法执行）。

```
within(com.iss.service.*)
```

（7）在 service 包或其子包中的任意连接点（在 Spring AOP 中只是方法执行）。

```
within(com.iss.service..*)
```

（8）实现了 AccountService 接口的代理对象的任意连接点（在 Spring AOP 中只是方法执行）。

```
this(com.iss.service.AccountService)
```

（9）实现 AccountService 接口的目标对象的任意连接点（在 Spring AOP 中只是方法执行）。

```
target(com.iss.service.AccountService)
```

（10）任何一个只接受一个参数，并且运行时所传入的参数是 Serializable 接口的连接点（在 Spring AOP 中只是方法执行）。

```
args(java.io.Serializable)
```

（11）请注意在例子中给出的切入点不同于 execution(* *(java.io.Serializable))：args 版本只有在动态运行时候传入参数是 Serializable 时才匹配，而 execution 版本在方法签名中声明只有一个 Serializable 类型的参数时候匹配。目标对象中有一个 @Transactional 注解的任意连接点（在 Spring AOP 中只是方法执行）。

```
@target(org.springframework.transaction.annotation.Transactional)
```

（12）任何一个目标对象声明的类型有一个 @Transactional 注解的连接点（在 Spring AOP 中只是方法执行）。

```
@within(org.springframework.transaction.annotation.Transactional)
```

（13）任何一个执行的方法有一个 @Transactional 注解的连接点（在 Spring AOP 中只是方法执行）。

```
@annotation(org.springframework.transaction.annotation.Transactional)
```

（14）任何一个只接受一个参数，并且运行时所传入的参数类型具有 @Classified 注解的连接点（在 Spring AOP 中只是方法执行）。

```
@args(com.iss.security.Classified)
```

（15）任何一个在名为 'tradeService' 的 Spring bean 之上的连接点（在 Spring AOP 中只是方法执行）。

```
bean(tradeService)
```

（16）任何一个在名字匹配通配符表达式 '*Service' 的 Spring bean 之上的连接点（在 Spring AOP 中只是方法执行）。

```
bean(*Service)
```

4.4 使用 AspectJ 实现注解增强

专家讲解

（1）AspectJ 是一个面向切面的框架，它扩展了 Java 语言，定义了 AOP 语法，能够在编译期提供代码的织入。

（2）@AspectJ 是 AspectJ5 新增的功能，使用 JDK5.0 注解技术和正规的 AspectJ 切点表达式语言描述切面。

（3）Spring 通过集成 AspectJ 实现了以注解的方式定义增强类，大大减少了配置文件中的工作量。

（4）使用 @AspectJ，首先要保证所用的是 JDK1.5 或以上版本。目前该项基本都能保证，公司一般使用 JDK1.8 版本。

（5）将 Spring 的 asm 模块添加到类路径中，以处理 @AspectJ 中所描述的方法参数名。

4.4.1 定义前置增强和后置增强

在 Maven 项目中添加 Spring AOP 相关的 jar 文件，使用注解定义前置增强和后置增强实现日志记录操作功能。

1）织入注解定义的增强

```
@Aspect
public class UserBizLogger {
private static final Logger log = Logger.getLogger(UserBizLogger.class);
// 定义前置增强
@Before("execution(* biz.IUserBiz.*(..))")
public void before( ) { … }
// 定义后置增强
@AfterReturning("execution(* biz.IUserBiz.*(..))")
public void afterReturning( ) { … }
}
```

2）编写 Spring 配置文件

```
<bean id="dao" class="dao.impl.UserDao"></bean>
<bean id="biz" class="biz.impl.UserBiz">
<property name="dao" ref="dao"></property>
</bean>
<!– 定义包含注解的增强类的实例 -->
<bean class="aop.UserBizLogger"></bean>
<!-- 织入使用注解定义的增强，需要引入 aop 命名空间 -->
<aop:aspectj-autoproxy />
```

4.4.2 获取连接点信息

如果想要获得被代理对象及方法的相关信息，又该怎么做呢？看看下面的示例，使用 JoinPoint 类。

```
@Aspect
public class UserBizLogger {
private static final Logger log = Logger.getLogger(UserBizLogger.class);
@Before("execution(* biz.IUserBiz.*(..))")
public void before(JoinPoint jp) {
        log.info(" 调用 " + jp.getTarget( ) + " 的 " + jp.getSignature( ).getName( )
        + " 方法。方法入参：" + Arrays.toString(jp.getArgs( )));
```

```
}
@AfterReturning(pointcut = "execution(* biz.IUserBiz.*(..))", returning = "returnValue")
public void afterReturning(JoinPoint jp, Object  returnValue) {
        log.info(" 调用 " + jp.getTarget( ) + " 的 " + jp.getSignature( ).getName( )
                + " 方法。方法返回值: " + returnValue);
}
}
```

4.4.3 定义最终增强

@AspectJ 还提供了一种最终增强类型，无论方法抛出异常还是正常退出，该增强都会得到执行，类似于异常处理机制中 finally 块的作用，一般用于释放资源，具体代码内容如下。

```
@Aspect
public class AfterLogger {
private static final Logger log = Logger.getLogger(AfterLogger.class);

@After("execution(* biz.IUserBiz.*(..))")
public void afterLogger(JoinPoint jp) {
        log.info(jp.getSignature( ).getName( ) + " 方法结束执行。");
}
}
```

4.4.4 定义异常抛出增强

```
@Aspect
public class ErrorLogger {
private static final Logger log = Logger.getLogger(ErrorLogger.class);

@AfterThrowing(pointcut = "execution(* biz.IUserBiz.*(..))", throwing = "e")
public void afterThrowing(JoinPoint jp, RuntimeException e) {
        log.error(jp.getSignature( ).getName( ) + " 方法发生异常: " + e);
}
}
```

4.4.5 定义环绕增强

ProceedingJoinPoint 是 JoinPoint 的子接口，它的 proceed( ) 方法可以调用真正的目标方法。

```
@Aspect
public class AroundLogger {
    private static final Logger log = Logger.getLogger(AroundLogger.class);
    @Around("execution(* biz.IUserBiz.*(..))")
    public Object aroundLogger(ProceedingJoinPoint jp) throws Throwable {
        log.info(" 调用 " + jp.getTarget( ) + " 的 " + jp.getSignature( ).getName( )
            + " 方法。方法入参：" + Arrays.toString(jp.getArgs( )));
        try {  Object result = jp.proceed( );
            log.info(" 调用 " + jp.getTarget( ) + " 的 "
        + jp.getSignature( ).getName( ) + " 方法。方法返回值：" + result);
            return result;
        } catch (Throwable e) {
    log.error(jp.getSignature( ).getName( ) + " 方法发生异常：" + e);
    throw e; }
    }
}
```

4.5 综合实例:猴子偷桃

引用一个猴子偷桃，守护者守护果园抓住猴子的小情节演示 Spring AOP。

1）猴子偷桃类（普通类）

```
package com.samter.common;
public class Monkey {
    public void stealPeaches(String name){
      System.out.println("【猴子】"+name+" 正在偷桃 ...");
  }
}
```

2）守护者类（声明为 Aspect）

```
package com.samter.aspect;
/**
 * 桃园守护者
 */
@Aspect
public class Guardian {
```

```
@Pointcut("execution(* com.samter.common.Monkey.stealPeaches(..))")
   public void foundMonkey( ){}
   @Before(value="foundMonkey( )")
   public void foundBefore( ){
      System.out.println("【守护者】发现猴子正在进入果园 ...");
   }
   @AfterReturning("foundMonkey( ) && args(name,..)")
   public void foundAfter(String name){
System.out.println("【守护者】抓住猴子，审问出猴子的名字叫“"+name+"”");
   }
}
```

3）XML 配置文件

```
<!-- 定义 Aspect -->
<bean id="guardian" class="com.samter.aspect.Guardian" />
<!-- 定义 Common -->
<bean id="monkey" class="com.samter.common.Monkey" />
<!-- 启动 AspectJ 支持 -->
<aop:aspectj-autoproxy />
```

4）编写测试类

```
public class Main {
public static void main(String[] args) {
ApplicationContext context = new ClassPathXmlApplicationContext
 ("config.xml");   Monkey monkey = (Monkey) context.getBean("monkey");
 try {
   monkey.stealPeaches(" 孙大圣的大徒弟 ");
    }
  catch(Exception e) {}
   }
}
```

5）测试运行输出结果

```
【守护者】发现猴子正在进入果园 ...
【猴子】孙大圣的大徒弟正在偷桃 ...
【守护者】抓住了猴子，守护者审问出了猴子的名字叫“孙大圣的大徒弟”...
```

专家讲解

本例写了一个猴子偷桃的方法。写了一个标志为 @Aspect 的类，它是守护者。它会在猴子偷桃之前发现猴子，并在猴子偷桃之后抓住猴子。

原理：

（1）@Aspect 的声明表示这是一个切面类。

（2）@Pointcut 使用这个方法可以将 com.samter.common.Monkey.stealPeaches(..) 方法声明为 poincut 即切入点。作用在 stealPeaches 方法被调用的时候执行守护类的 foundMonkey 方法。其中 execution 是匹配方法执行的切入点，也就是 spring 最常用的切入点定义方式。

（3）@Before(value="foundMonkey( )")：@Before 声明为在切入点方法执行之前执行，而后面没有直接声明切入点，而是 value="foundMonkey( )"，是因为如果 @afterReturning 等都有所改动的时候都必须全部改动，所以统一用 Pointcut 的 foundMonkey 代替，这样有改动的时候仅需改动一个地方。其他 @AfterReturning 类同。

小结

（1）Spring 提供的增强处理类型包括：前置增强、后置增强、环绕增强、异常抛出增强、最终增强等。

（2）使用注解方式定义切面可以简化配置工作量，通过在配置文件中添加 <aop-:aspectj-autoproxy /> 元素启用对于 @AspectJ 注解的支持。

（3）常用的注解有 @Aspect、@Before、@AfterReturning、@Around、@AfterThrowing、@After 等。

（4）也可以通过 Schema 形式将 POJO 的方法配置成切面，所用标签包括 <aop:aspect>、<aop:before>、<aop:after-returning>、<aop:around>、<aop:after-throwing>、<aop:after> 等。

经典面试题

（1）什么是 AOP?

（2）为什么要使用 AOP?

（3）Spring AOP 能够在什么场景下使用？

（4）解释 Aspect 切面？

（5）在 Spring AOP 中，关注点和横切关注的区别是什么？

（6）什么是连接点？

（7）什么是通知？

（8）什么是切点？

（9）什么是基于注解的切面实现？

（10）怎样开启注解装配？

跟我上机

（1）使用 Spring AOP 实现系统登录日志功能。

①业务介绍：将业务逻辑层方法的调用信息输出到控制台。

② AOP 思路：分别编写业务逻辑代码和“增强”代码，运行时再组装。

③使用前置增强和后置增强对业务方法的执行过程进行日志记录。

（2）使用 @AspectJ 注解实现注册功能日志增强。

使用注解方式定义前置增强（包含连接点信息）和后置增强（包含连接点信息和返回值），对业务方法的执行过程进行日志记录。

第 5 章　Spring JDBC 框架

本章要点(学会后请在方框中打钩):

- □　了解 Spring JDBC 框架
- □　使用 JdbcTemplate 进行数据库 CRUD 操作
- □　了解异常转换
- □　使用 Simple JDBC 类实现 JDBC 操作

5.1 解释 Spring JDBC 框架

Spring JDBC 模块是 Spring 框架的基础模块之一，提供了一套 JDBC 抽象框架，用于简化 JDBC 开发，如图 1 所示。

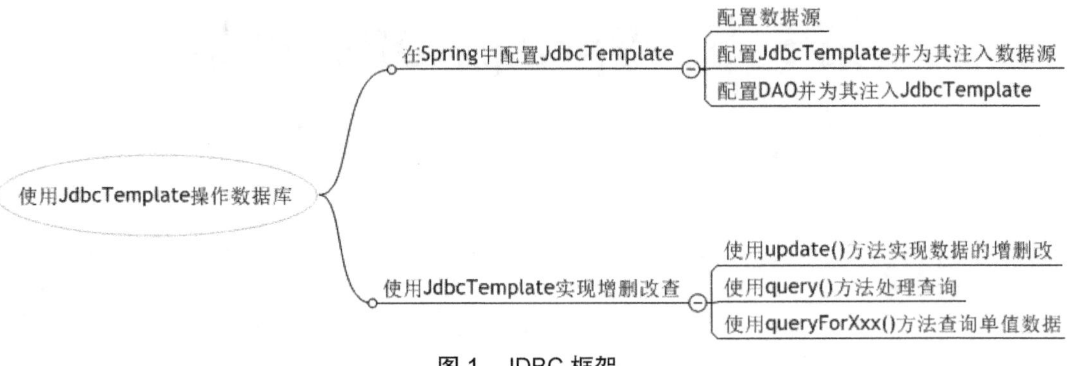

图 1　JDBC 框架

5.1.1 Spring JDBC 概述

（1）Spring JDBC 抽象框架 core 包提供了 JDBC 模板类，其中 JdbcTemplate 是 core 包的核心类，所以其他模板类都是基于它封装完成的，JDBC 模板类是第一种工作模式。

（2）JdbcTemplate 类通过模板设计模式帮助我们消除了冗长的代码，只做需要做的事情（即可变部分），并且帮我们做那些固定部分，如连接的创建及关闭。

（3）JdbcTemplate 类对可变部分采用回调接口方式实现，如 ConnectionCallback 通过回调接口返回给用户一个连接，从而可以使用该连接做任何事情、StatementCallback 通过回调接口返回给用户一个 Statement，从而可以使用该 Statement 做任何事情，等等，还有其他一些回调接口如图 2 所示。

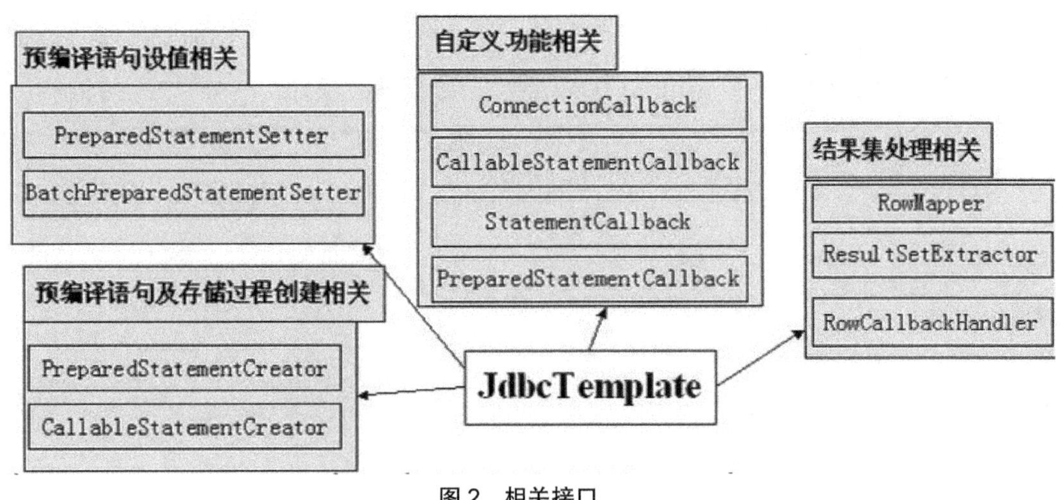

图 2　相关接口

（4）JdbcTemplate 支持的回调接口。

（5）Spring 除了提供 JdbcTemplate 核心类，还提供了基于 JdbcTemplate 实现的 NamedParameterJdbcTemplate 类用于支持命名参数绑定、SimpleJdbcTemplate 类用于支持 Java5+ 的可变参数及自动装箱拆箱等特性。

专家讲解

在 Spring JDBC 模块中，所有的类可以被分到四个单独的包。

（1）core：即核心包，它包含了 JDBC 的核心功能。此包内有很多重要的类，包括：JdbcTemplate 类、SimpleJdbcInsert 类，SimpleJdbcCall 类，以及NamedParameterJdbcTemplate 类。

（2）datasource：即数据源包，访问数据源的实用工具类。它有多种数据源的实现，可以在 Java EE 容器外部测试 JDBC 代码。

（3）object：即对象包，以面向对象的方式访问数据库。它允许执行查询并返回结果作为业务对象。它可以在数据表的列和业务对象的属性之间映射查询结果。

（4）support：即支持包，是 core 包和 object 包的支持类。例如提供了异常转换功能的 SQLException 类。

5.1.2 注解配置数据源

下面以 MySQL 数据库为例，开始简单的数据源配置。

```
@Configuration
@ComponentScan("com.ch.myalbumjdbc")
public class SpringJdbcConfig {
    @Bean
    public DataSource mysqlDataSource( ) {
    DriverManagerDataSource dataSource = new DriverManagerDataSource( );
    dataSource.setDriverClassName("com.mysql.jdbc.Driver");
    dataSource.setUrl("jdbc:mysql://localhost:3306/springjdbc");
    dataSource.setUsername(" 数据库用户名 ");
    dataSource.setPassword(" 数据库密码 ");
    return dataSource;
  }
}
```

5.1.3 XML 配置数据源

```
<bean id="dataSource" class="org.apache.commons.dbcp.BasicDataSource"
 destroy-method="close">
   <property name="driverClassName" value="com.mysql.jdbc.Driver"/>
  <property name="url" value="jdbc:mysql://localhost:3306/springjdbc"/>
  <property name="username" value=" 数据库用户名 "/>
  <property name="password" value=" 数据库密码 "/>
</bean>
```

5.2 传统 JDBC 编程替代方案

5.2.1 JdbcTemplate 基本的查询

JDBC 模板是 Spring JDBC 模块中主要的 API，它提供了常见的数据库访问功能。

```
int result = jdbcTemplate.queryForObject("SELECT COUNT(*) FROM EMPLOYEE", Integer.class);
```

5.2.2 JdbcTemplate 简单的插入功能

```
public int addEmplyee(int id) {
String sql="INSERT INTO EMPLOYEE VALUES (?,?,?,?)";
return jdbcTemplate.update(sql,5,"Bill","Gates","USA");
}
```

5.2.3 JdbcTemplate 行映射——RowMapper

还有一个非常有用的功能是把查询结果映射到 Java 对象——通过实现 RowMapper 接口。

例如，对于查询返回的每一行结果，Spring 会使用该行映射 (RowMapper) 来填充 Java bean。

```
public class EmployeeRowMapper implements RowMapper<Employee> {
    @Override
  public Employee mapRow(ResultSet rs, int rowNum) throws SQLException {
      Employee employee = new Employee( );
      employee.setId(rs.getInt("ID"));
```

```
        employee.setFirstName(rs.getString("FIRST_NAME"));
        employee.setLastName(rs.getString("LAST_NAME"));
        employee.setAddress(rs.getString("ADDRESS"));
         return employee;
    }
}
```

现在，我们传递行映射器给查询 API，并得到完全填充好的 Java 对象。

```
String query = "SELECT * FROM EMPLOYEE WHERE ID = ?";
List<Employee> employees = jdbcTemplate.queryForObject( query, new Object[] { id }, new EmployeeRowMapper( ));
```

5.3 异常转换

Spring 提供了自己的开箱即用的数据异常分层——DataAccessException 作为根异常，它负责转换所有的原始异常。

所以开发者无须处理底层的持久化异常，因为 Spring JDBC 模块已经在 DataAccessException 类及其子类中封装了底层的异常。这样可以使异常处理机制独立于当前使用的具体数据库。

除了默认的 SQLErrorCodeSQLExceptionTranslator 类，开发者也可以提供自己的 SQLExceptionTranslator 实现。

下面是一个自定义 SQLExceptionTranslator 实现的简单例子，当出现完整性约束错误时自定义错误消息。

```
public class CustomSQLErrorCodeTranslator extends SQLErrorCodeSQLExceptionTranslator {
    @Override
    protected DataAccessException customTranslate(String task, String sql, SQLException sqlException) {
    if (sqlException.getErrorCode( ) == -104) {
        return new DuplicateKeyException( "Custom Exception translator - Integrity constraint violation.", sqlException);
    }
    return null;
    }
}
```

要使用自定义的异常转换器，我们需要把它传递给 JdbcTemplate——通过 callingsetExceptionTranslator( ) 方法。

```
CustomSQLErrorCodeTranslator customSQLErrorCodeTranslator = new CustomSQLErrorCodeTranslator( );
jdbcTemplate.setExceptionTranslator(customSQLErrorCodeTranslator);
```

5.4 使用 SimpleJdbc 类实现 JDBC 操作

SimpleJdbc 类提供简单的方法来配置和执行 SQL 语句。这些类使用数据库的元数据来构建基本的查询。 SimpleJdbcInsert 类和 SimpleJdbcCall 类提供了更简单的方式来执行插入和存储过程的调用。

5.4.1 SimpleJdbcInsert 类

下面来看看执行简单的插入语句的最低配置，基于 SimpleJdbcInsert 类的配置产生的 INSERT 语句。

所有需要提供的是：表名、列名和值。让我们先创建 SimpleJdbcInsert。

```
SimpleJdbcInsert simpleJdbcInsert = new SimpleJdbcInsert(dataSource).withTableName("EMPLOYEE");
```

现在，让我们提供列名和值，并执行操作。

```
public int addEmplyee(Employee emp) {
   Map<String, Object> parameters = new HashMap<String, Object>( );
   parameters.put("ID", emp.getId( ));
   parameters.put("FIRST_NAME", emp.getFirstName( ));
   parameters.put("LAST_NAME", emp.getLastName( ));
   parameters.put("ADDRESS", emp.getAddress( ));
   return simpleJdbcInsert.execute(parameters);
}
```

为了让数据库生成主键，可以使用 executeAndReturnKey( ) API，我们还需要配置的实际自动生成的列。

```
SimpleJdbcInsert simpleJdbcInsert = new SimpleJdbcInsert(dataSource)
      .withTableName("EMPLOYEE")
.usingGeneratedKeyColumns("ID");
 Number id = simpleJdbcInsert.executeAndReturnKey(parameters);
   System.out.println("Generated id - " + id.longValue( ));
```

最后还能使用 BeanPropertySqlParameterSource 和 MapSqlParameterSource 传递数据。

5.4.2 用 SimpleJdbcCall 调用存储过程

让我们看看如何执行存储过程——使用 SimpleJdbcCall 的抽象。

```
SimpleJdbcCall simpleJdbcCall = new SimpleJdbcCall(dataSource) .withProcedureName("READ_EMPLOYEE");
public Employee getEmployeeUsingSimpleJdbcCall(int id) {
SqlParameterSource in = new MapSqlParameterSource( ).addValue("in_id", id);  Map<String, Object> out = simpleJdbcCall.execute(in);
Employee emp = new Employee( );
emp.setFirstName((String) out.get("FIRST_NAME"));
emp.setLastName((String) out.get("LAST_NAME"));
return emp;
}
```

5.4.3 批处理操作

另一个简单的例子——把多种操作合在一起实现批处理。

1）使用 JdbcTemplate 执行基本的批处理操作

使用 JdbcTemplate 类，通过 batchUpdate( ) API 来执行基本的批处理操作，注意，BatchPreparedStatementSetter 实现是很有趣的。

```
public int[] batchUpdateUsingJdbcTemplate(List<Employee> employees) {
return jdbcTemplate.batchUpdate("INSERT INTO EMPLOYEE VALUES (?, ?, ?, ?)", new BatchPreparedStatementSetter( ) {
@Override
public void setValues(PreparedStatement ps, int i) throws SQLException {
ps.setInt(1, employees.get(i).getId( ));
ps.setString(2, employees.get(i).getFirstName( ));
ps.setString(3, employees.get(i).getLastName( ));
ps.setString(4, employees.get(i).getAddress( );
}
@Override
public int getBatchSize( ) {
return 50;
}
});
}
```

2）使用 NamedParameterJdbcTemplate 执行批处理操作

对于批处理操作，还可以选择使用 NamedParameterJdbcTemplate 的 batchUpdate( ) API 来执行。

此 API 比先前的更简单——无须实现任何额外的接口来设置参数，因为它有一个内部的预准备语句的 setter 来传递预设的参数值。参数值可以通过 batchUpdate( ) 方法传递给 SqlParameterSource 的数组。

```
SqlParameterSource[] batch=SqlParameterSourceUtils.createBatch(employees.toArray( ));
int[] updateCounts = namedParameterJdbcTemplate.batchUpdate(
"INSERT INTO EMPLOYEE VALUES (:id, :firstName, :lastName, :address)", batch);
 return updateCounts;
```

小结

（1）Spring JDBC 提供了一套 JDBC 抽象框架，用于简化 JDBC 开发。

（2）Spring 主要提供 JDBC 模板方式、关系数据库对象化方式、SimpleJdbc 方式、事务管理来简化 JDBC 编程。

（3）Spring 提供了 3 个模板类。

① JdbcTemplate：Spring 里最基本的 JDBC 模板，利用 JDBC 和简单的索引参数查询提供对数据库的简单访问。

② NamedParameterJdbcTemplate：能够在执行查询时把值绑定到 SQL 里的命名参数，而不是使用索引参数。

③ SimpleJdbcTemplate：利用 Java5 的特性，比如自动装箱、通用（generic）和可变参数列表来简化 JDBC 模板的使用。

（4）JdbcTemplate 主要提供以下 4 类方法。

① execute 方法：可以用于执行任何 SQL 语句，一般用于执行 DDL 语句。

② update 方法及 batchUpdate 方法：update 方法用于执行新增、修改、删除等语句；batchUpdate 方法用于执行批处理相关语句。

③ query 方法及 queryForXXX 方法：用于执行查询相关语句。

④ call 方法：用于执行存储过程、函数相关语句。

经典面试题

（1）spring 的 jdbc 与传统的 jdbc 有什么区别？

（2）在 Web 项目中如何配置 Spring?

（3）在 Spring 框架中如何更有效地使用 JDBC?

（4）解释什么是 JdbcTemplate。

（5）列举 Spring 支持的 ORM 框架有哪些。

跟我上机

（1）建立 Web 项目，使用 Spring+JdbcTemplate 编写一个缩写词查询系统。
要求能够根据缩写词查询缩写词的全拼单词组合。
已知 MyAccronym 数据库中 Accroyms 表字段信息如表 1 所示。

表 1 字段信息

数据库名	MyAccronym		表名	Accroyms
字段显示	字段名	数据类型	字段大小	备注和说明
缩写词编号	Id	int	1	标识列，主键
缩写	ShortForm	varchar	50	
完整解释	LongForm	varchar	500	

Accroyms 表参考数据信息，如图 3 所示。

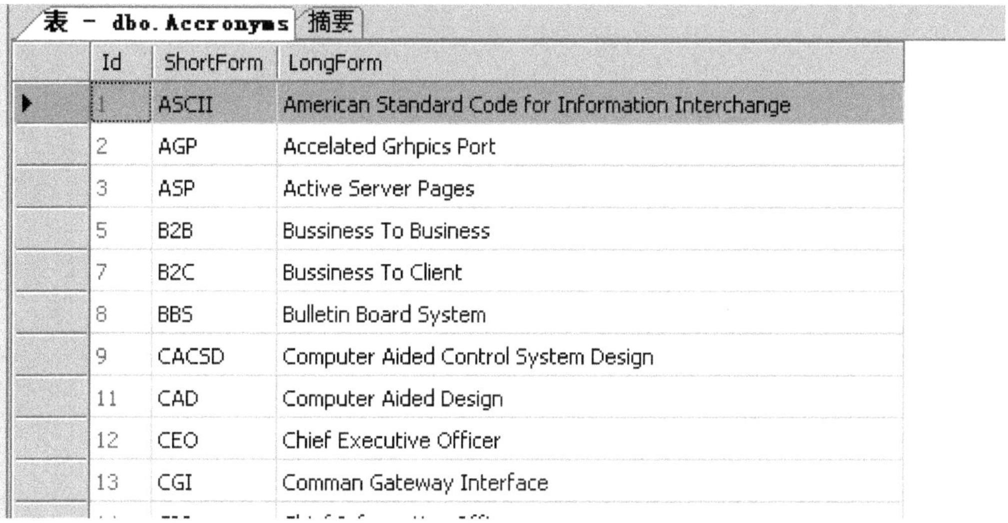

表 - dbo.Accronyms 摘要

Id	ShortForm	LongForm
1	ASCII	American Standard Code for Information Interchange
2	AGP	Accelated Grhpics Port
3	ASP	Active Server Pages
5	B2B	Bussiness To Business
7	B2C	Bussiness To Client
8	BBS	Bulletin Board System
9	CACSD	Computer Aided Control System Design
11	CAD	Computer Aided Design
12	CEO	Chief Executive Officer
13	CGI	Comman Gateway Interface

图 3 Accroyms 表参考数据信息

第 6 章　Spring 事务管理

本章要点（学会后请在方框中打钩）：

- □ 了解什么是事务
- □ 了解事务的特性
- □ 掌握 Spring 编程式事务
- □ 掌握 Spring 声明式事务

6.1 什么是事务

这里的事务的概念，最好在计算机环境下来理解。在操作系统中，我们为了保证一段程序执行过程不被中断，而引入了互斥锁的概念。这里事务的概念可以看作对于互斥锁的事件过程的另一种解释。即，一个事务执行的前后，数据资源的状态始终是正确的。由此，我们得到了事务的几个基本属性：原子性，一致性，隔离性，持久性。

6.1.1 原子性

原子性要求事务所包含的全部操作是一个不可分割的整体，这些操作要么全部提交成功，只要其中一个操作失败，就全部失败。

6.1.2 一致性

一致性要求事务在执行之前数据资源是保持一致的，在执行之后数据资源仍然保持一致。举个例子来理解：A，B 两个变量之和是 10 为大前提，对其中一个变量改变值之后的结果，必须也要满足这个大前提。银行里的转账操作就是一致性的非常好的例子。

6.1.3 隔离性

隔离性即规定了不同事务之间的相互影响程度，换句话说就是不同事务之间的执行时机。这里的隔离性就衍生出原子性、一致性的概念。当一个事务开始执行对数据资源的访问时，另一个事务合适能够访问数据资源就决定了隔离性的大小。这里分为 4 个级别，由宽松到严格的顺序为：“Read Uncommitted”，“Read Committed”，“Repeatable Read”和“Serializable”。

6.1.4 持久性

持久性要求事务对于数据资源的操作不可逆转，即事务完成之后，程序执行流已经彻底退出该事务，该事务对应的程序执行流不可回退。通俗地讲，就是“不要赖账”。

6.2 Spring 编程式事务

我们先来介绍 Spring 中的第一种方式：使用“编程式事务管理”的实现，本例采用 Spring+MyBatis 框架整合开发实现，如果不能理解请学完 MyBatis 后再回来看本章。

建立 Spring6-1_ 编程式事务，工程基本结构图略。

（1）创建 BankDao.java 文件，具体内容如下。

```
package com.iss.dao;
import com.iss.entity.Account;
public interface BankDao {
    public void  update(Account account);
}
```

（2）创建 Account.java 文件，具体内容如下。

```
public class Account {
    private int id;
    private int money;
    // 省略 setter/getter
    public Account(int id, int money) {
        super( );
        this.id = id;
        this.money = money;
    }
    public Account( ) {
        super( );
    }
}
```

（3）创建 IBankService.java 文件，具体内容如下。

```
public interface IBankService {
    public void transferMoney(int sourceId,int desId,int money);
}
```

（4）创建 BankServiceImpl.java 文件，具体内容如下。

```
@Service
public class BankServiceImpl implements IBankService{
    @Autowired
    private BankDao bankDao;
    private TransactionTemplate transactionTemplate;
     public void setBankDao(BankDao bankDao) {
        this.bankDao = bankDao;
    }
    public void setTransactionTemplate(TransactionTemplate transactionTemplate) {
        this.transactionTemplate = transactionTemplate;
```

```
    }
    public void transferMoney(final int countA,final int countB,final int money) {
        transactionTemplate.execute(new TransactionCallbackWithoutResult( ) {
            @Override
            protected void doInTransactionWithoutResult(TransactionStatus status) {
                Account accountA = new Account(countA,money);
                Account accountB = new Account(countB,money);
                bankDao.updateOut(accountA);
                //System.out.println(1/0);
                bankDao.updateIn(accountB);
            }
        });
    }
}
```

（5）创建 beans.xml 文件，具体内容如下。

```
<context:component-scan base-package="com.iss" />
    <bean id="dataSource" class="org.apache.commons.dbcp.BasicDataSource"
        destroy-method="close">
        <property name="driverClassName" value="com.mysql.jdbc.Driver" />
        <property name="url" value="jdbc:mysql://localhost:3306/db_bank" />
        <property name="username" value="root" />
        <property name="password" value="root" />
    </bean>
    <!-- 配置 mybatis 的 sqlSessionFactory -->
   <bean id="sqlSessionFactory" class="org.mybatis.spring.SqlSessionFactoryBean">
       <property name="dataSource" ref="dataSource" />
       <!-- 自动扫描 mappers.xml 文件 -->
    <property name="mapperLocations" value="classpath:mappers/*.xml"></property>
   </bean>
   <!-- DAO 接口所在包名，Spring 会自动查找其下的类 -->
   <bean class="org.mybatis.spring.mapper.MapperScannerConfigurer">
       <property name="basePackage" value="com.iss.dao" />
       <property name="sqlSessionFactoryBeanName" value="sqlSessionFactory">
       </property>
   </bean>
   <!-- jdbc 事务管理器 -->
```

```
    <bean id="transactionManager"
class="org.springframework.jdbc.datasource.DataSourceTransactionManager">
        <property name="dataSource" ref="dataSource"></property>
    </bean>

    <bean id="transactionTemplate"
        class="org.springframework.transaction.support.TransactionTemplate">
        <constructor-arg
            type="org.springframework.transaction.PlatformTransactionManager"
            ref="transactionManager" />
    </bean>
    <bean id="namedParameterJdbcTemplate"
class="org.springframework.jdbc.core.namedparam.NamedParameterJdbcTemplate">
        <constructor-arg ref="dataSource"></constructor-arg>
    </bean>

    <bean id="bankService" class="com.iss.service.impl.BankServiceImpl">
        <property name="transactionTemplate" ref="transactionTemplate">
        </property>
    </bean>
</beans>
```

（6）创建 BankMapper.xml 文件，具体内容如下。

```
<?xml version="1.0" encoding="UTF-8"?>
<!DOCTYPE mapper
PUBLIC "-//mybatis.org//DTD Mapper 3.0//EN"
"http://mybatis.org/dtd/mybatis-3-mapper.dtd">
<mapper namespace="com.iss.dao.BankDao">
        <update id="updateIn" parameterType="com.iss.entity.Account">
            UPDATE account SET money=money+#{money} WHERE id=#{id}
    </update>
    <update id="updateOut" parameterType="com.iss.entity.Account">
            UPDATE account SET money=money-#{money} WHERE id=#{id}
    </update>
</mapper>
```

（7）创建 T.java 文件，具体内容如下。

```
public class T {
    public static void main(String[] args) {
    ApplicationContext context = new
ClassPathXmlApplicationContext("classpath:beans.xml");
    IBankService bankService = (IBankService) context.getBean("bankService");
        bankService.transferMoney(1,2,200);
    }
}
```

（8）创建数据库 db_bank 及 t_account 表（图 1）。

图 1　t_account 表

（9）运行测试结果，刷新数据库，观察数据库变化（图 2）。

图 2　刷新 t_account 表

（10）打开 System.out.println(1/0); 注释，再次运行，观察数据库变化。

6.3　Spring 声明式事务

上一节，我们演示了通过编程式方式进行事务管理。我们发现这种配置方式破坏了我们倡导的 AOP 切面编程，因此，我们来看一下如何通过声明的方式来实现 Spring 事务管理，来达到我们 AOP 切面的目的。本节内容使用了 Spring AOP 的部分内容，如果对此有疑问，再参考前面的叙述及实例。

复制 Spring6-1_ 编程式事务，重命名为 Spring6-2_ 声明式事务。

（1）修改 BankServiceImpl.java 文件，具体内容如下。

```
package com.iss.service.impl;
@Service
public class BankServiceImpl implements IBankService {
        @Autowired
        private BankDao bankDao;
        public void setBankDao(BankDao bankDao) {
                this.bankDao = bankDao;
        }
        public void transferMoney(int countA, int countB, int money) {
                Account accountA = new Account(countA, money);
                Account accountB = new Account(countB, money);
                bankDao.updateOut(accountA);
                // System.out.println(1/0);
                bankDao.updateIn(accountB);
        }
}
```

（2）修改 beans.xml 文件，具体内容如下。

```
// 命名空间略
        <context:component-scan base-package="com.iss" />
        <bean id="dataSource" class="org.apache.commons.dbcp.BasicDataSource"
                destroy-method="close">
                <property name="driverClassName" value="com.mysql.jdbc.Driver" />
                <property name="url" value="jdbc:mysql://localhost:3306/db_bank" />
                <property name="username" value="root" />
                <property name="password" value="root" />
        </bean>
        <!-- 配置 mybatis 的 sqlSessionFactory -->
        <bean id="sqlSessionFactory" class="org.mybatis.spring.SqlSessionFactoryBean">
                <property name="dataSource" ref="dataSource" />
                <!-- 自动扫描 mappers.xml 文件 -->
                <property name="mapperLocations" value="classpath:mappers/*.xml"></property>
        </bean>
```

```
        <!-- DAO 接口所在包名,Spring 会自动查找其下的类 -->
        <bean class="org.mybatis.spring.mapper.MapperScannerConfigurer">
            <property name="basePackage" value="com.iss.dao" />
            <property name="sqlSessionFactoryBeanName" value="sqlSessionFactory"></property>
        </bean>
        <!-- jdbc 事务管理器 -->
        <bean id="transactionManager"
            class="org.springframework.jdbc.datasource.DataSourceTransactionManager">
            <property name="dataSource" ref="dataSource"></property>
        </bean>
        <bean id="namedParameterJdbcTemplate"
          class="org.springframework.jdbc.core.namedparam.NamedParameterJdbcTemplate">
            <constructor-arg ref="dataSource"></constructor-arg>
        </bean>
        <bean id="bankService" class="com.iss.service.impl.BankServiceImpl">
        </bean>
        <!-- 配置事务通知 -->
        <tx:advice id="txAdvice" transaction-manager="transactionManager">
            <tx:attributes>
                <tx:method name="*"/>
            </tx:attributes>
        </tx:advice>
        <!-- 配置事务切面 -->
        <aop:config>
            <aop:pointcut id="serviceMethod" expression="execution(* com.java.ingo.service.*.*(..))"/>
            <!-- 配置事务通知 -->
            <aop:advisor advice-ref="txAdvice" pointcut-ref="serviceMethod"/>
        </aop:config>
```

专家讲解

（1）增加事务通知 advice 命名空间。xmlns:tx="http://www.springframework.org/schema/tx。

（2）针对方法配置事务，<tx:method name="*"/> 表示所有方法都作为事务处理。

（3）配置事务切面，包括切点，及通知属性。

（4）测试方法：运行 main 方法，然后刷新数据库，观察数据库变化。

（5）打开 System.out.println(1/0); 注解，再次运行 main 方法，观察数据库变化。

（3）修改 BankServiceImpl.java 文件，具体内容如下。

```
package com.iss.service.impl;
@Transactional
public class BankServiceImpl implements IBankService {
    @Autowired
    private BankDao bankDao;

    public void setBankDao(BankDao bankDao) {
        this.bankDao = bankDao;
    }

    public void transferMoney(int countA, int countB, int money) {
        Account accountA = new Account(countA, money);
        Account accountB = new Account(countB, money);
        bankDao.updateOut(accountA);
        // System.out.println(1/0);
        bankDao.updateIn(accountB);
    }
}
```

（4）修改 beans.xml 文件，具体内容如下。

```
// 命名空间略
    <context:component-scan base-package="com.iss" />
    <bean id="dataSource" class="org.apache.commons.dbcp.BasicDataSource"
        destroy-method="close">
        <property name="driverClassName" value="com.mysql.jdbc.Driver" />
        <property name="url" value="jdbc:mysql://localhost:3306/db_bank" />
        <property name="username" value="root" />
        <property name="password" value="root" />
    </bean>
    <!-- 配置 mybatis 的 sqlSessionFactory -->
    <bean id="sqlSessionFactory" class="org.mybatis.spring.SqlSessionFactoryBean">
        <property name="dataSource" ref="dataSource" />
        <!-- 自动扫描 mappers.xml 文件 -->
        <property name="mapperLocations" value="classpath:mappers/*.xml"></property>
    </bean>
    <!-- DAO 接口所在包名，Spring 会自动查找其下的类 -->
```

```
    <bean class="org.mybatis.spring.mapper.MapperScannerConfigurer">
        <property name="basePackage" value="com.iss.dao" />
        <property name="sqlSessionFactoryBeanName"
value="sqlSessionFactory"></property>
    </bean>
    <!-- jdbc 事务管理器 -->
    <bean id="transactionManager"
     class="org.springframework.jdbc.datasource.DataSourceTransactionManager">
        <property name="dataSource" ref="dataSource"></property>
    </bean>
    <bean id="namedParameterJdbcTemplate"
class="org.springframework.jdbc.core.namedparam.NamedParameterJdbcTemplate">
        <constructor-arg ref="dataSource"></constructor-arg>
    </bean>
    <bean id="bankService" class="com.iss.service.impl.BankServiceImpl">
    </bean>
    <!-- 织入事务管理器 -->
    <tx:annotation-driven transaction-manager="transactionManager"/>
</beans>
```

（4）测试方法：运行 main 方法，然后刷新数据库，观察数据库变化。
（5）打开 System.out.println(1/0); 注解，再次运行，观察数据库变化。

小结

（1）我们发现编程式事务管理的方法配置内容过多，维护不易，易出错。
（2）本文第一种 xml 方法为推荐方式，配置范围易管理。
（3）本文第二种方法使用时最方便，但服务过多时，注解范围维护不易。

经典面试题

（1）什么是事务？
（2）列举事务的 4 种基本属性。
（3）Spring 支持的事务管理类型有几种？
（4）Spring 框架的事务管理有哪些优点？
（5）你更倾向于哪种事务管理类型？

跟我上机

模拟银行转账功能的实现，要求使用声明式事务控制实现。

例如：A—>B 转账，对应以下两条 sql 语句。

update account set money=money-100 where name= ‘a’；

update account set money=money+100 where name= ‘b’；

第 2 篇　Spring MVC 框架

学习目标

- □ 了解 Spring MVC 框架简介
- □ 精通 Spring MVC 框架的配置过程
- □ 熟悉 HTTP 请求地址映射
- □ 熟悉 HTTP 请求数据的绑定
- □ 熟悉数据模型控制
- □ 熟悉视图及解析器
- □ 精通 Spring MVC 拦截器
- □ 精通 Spring MVC 文件上传
- □ 精通 Spring MVC 的异常处理
- □ 精通 Spring MVC 和 MyBatis 整合
- □ 精通 Spring，Spring MVC 和 MyBatis 整合

第 1 章　Spring MVC 框架入门

本章要点（学会后请在方框中打钩）：

- □ 了解什么是 Spring MVC 框架
- □ 掌握 Spring MVC 框架的运行原理
- □ 掌握手动搭建 Spring MVC 的实现过程
- □ 掌握创建 Maven Web 项目的方法和过程

1.1 Spring MVC 介绍

Spring Web MVC 是一种基于 Java 的实现了 Web MVC 设计模式的请求驱动类型的轻量级 Web 框架,即使用 MVC 架构模式的思想,将 web 层进行职责解耦,基于请求驱动指的就是使用请求—响应模型,框架的目的就是帮助我们简化开发, Spring Web MVC 也是帮助简化我们日常 Web 开发的。

另外还有一种基于组件的、事件驱动的 Web 框架在此就不介绍了,如 Tapestry、JSF 等。

Spring Web MVC 也是服务到工作者模式的实现,但进行可优化。前端控制器是 DispatcherServlet;应用控制器其实拆为处理器映射器 (Handler Mapping) 进行处理器管理和视图解析器 (View Resolver) 进行视图管理;页面控制器 / 动作 / 处理器为 Controller 接口(仅包含 Model And View handle Request(request, response)方法)的实现(也可以是任何的 POJO 类);支持本地化(Locale)解析、主题(Theme)解析及文件上传等;提供了非常灵活的数据验证、格式化和数据绑定机制;提供了强大的约定大于配置(惯例优先原则)的契约式编程支持。

MVC 全名是 Model View Controller,是模型 (model) 一视图 (view) 一控制器 (controller) 的缩写,用一种业务逻辑、数据、界面显示分离的方法组织代码,能够将业务逻辑聚集到一个部件里面。

Spring MVC 是当前最优秀的 MVC 框架,自从 Spring 2.5 版本发布后,由于支持注解配置,易用性有了大幅度的提高。Spring 5.0 更加完善,已经实现了对 Struts2 的超越。现在越来越多的开发团队选择了 Spring MVC。

Struts2 也是非常优秀的 MVC 构架,优点非常多,比如良好的结构,拦截器的思想,丰富的功能。但这里想说的是缺点, Struts2 由于采用了值栈、OGNL 表达式、Struts2 标签库等,会导致应用的性能下降,应避免使用这些功能。而 Struts2 的多层拦截器、多实例 action 性能都很好。

1.2 Spring MVC 的优点

Spring MVC 使用简单,学习成本低。学习难度小于 Struts2, Struts2 用不上的多余功能太多,当然这不是决定因素。

Spring MVC 很容易就可以写出性能优秀的程序,Struts2 要处处小心才可以写出性能优秀的程序(指 MVC 部分)。

Spring MVC 的灵活是你无法想像的,Spring 框架的扩展性有口皆碑,Spring MVC 当然也不会落后,不会因使用了 MVC 框架而感到有任何的限制。

Spring MVC 属于 Spring Frame Work 的后续产品,已经融合在 Spring Web Flow 里面。Spring 框架提供了构建 Web 应用程序的全功能 MVC 模块。使用 Spring 可插入的 MVC 架构,从而在使用 Spring 进行 Web 开发时,可以选择使用 Spring 的 Spring MVC 框架或集成其他 MVC 开发框架，如 Struts1，Struts2 等。

Spring MVC 框架是有一个 MVC 框架，通过实现 Model-View-Controller 模式来很好地将数据、业务与展现进行分离。从这个角度来说，Spring MVC 和 Struts1、Struts2 非常类似。Spring MVC 的设计是围绕 DispatcherServlet 展开的，DispatcherServlet 负责将请求派发到特定的 handler。通过可配置的 handler mappings、view resolution。

1.3 Spring MVC 运行原理

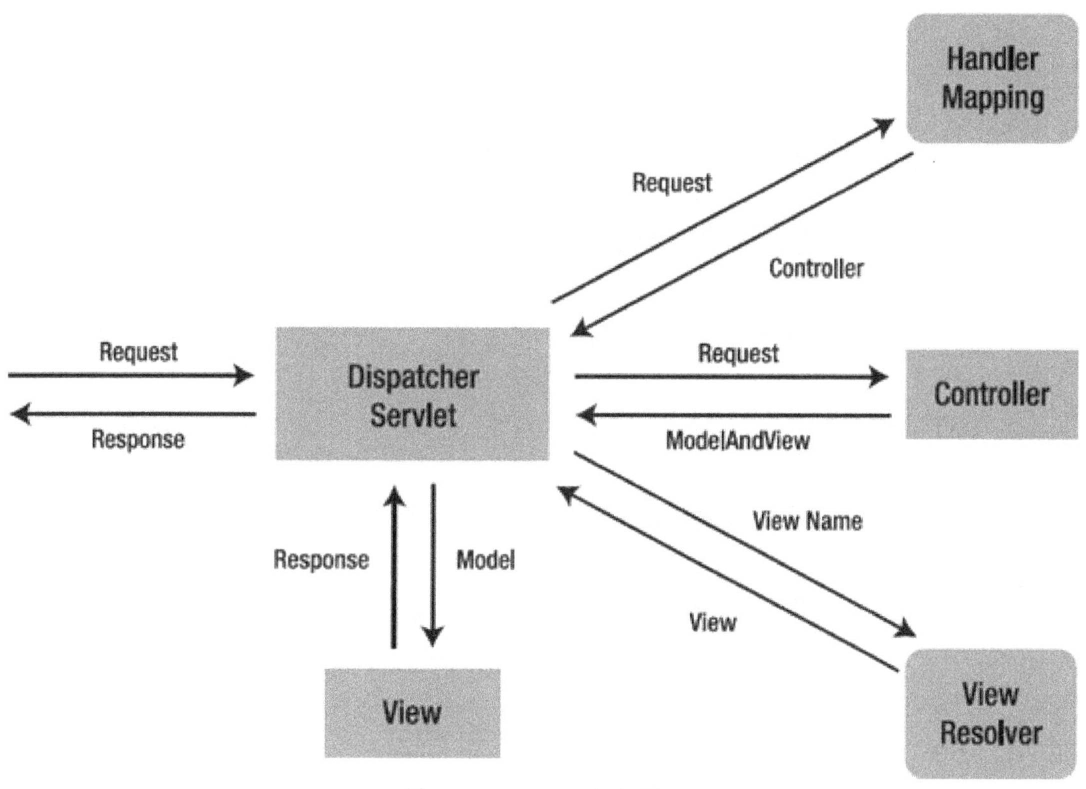

图 1 Spring MVC 运行原理

上面的示意图翻译如下。

（1）用户发送请求（Incoming reuqest）。

（2）前端控制器接受请求（Font controller）。前端控制器会自动选择出于当前请求路径，方法相匹配的 *controller 或者方法来接受当前请求。

（3）相匹配的 *controller，即实际处理当前请求的 Controller，调用业务对象、服务、工具等，将该请求的处理结果，通过 model 模型的方式交还给前端控制器（Font controller）

（4）前端控制器，根据视图层的规则，选择一个具体的 View template 并且进行渲染。

（5）将控制权交还给前端控制器。

（6）用户界面产生响应。

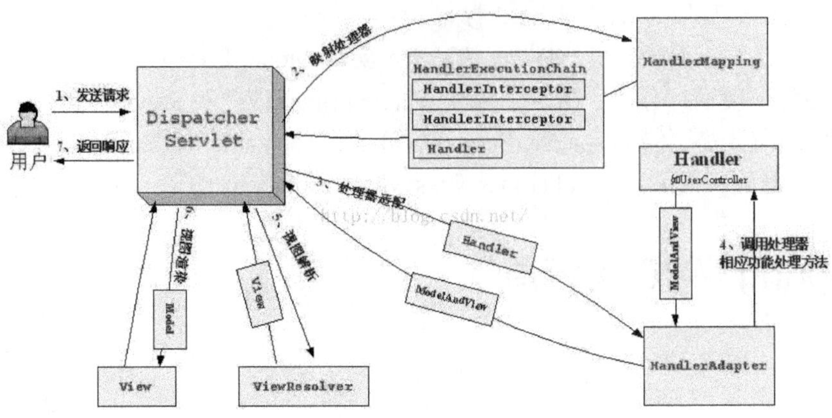

图 2 Spring MVC 运行流程

观察这两张图，我们可以得到如下结论。

（1）前端控制器，即 Dispatcher Servlet。它根据 HandlerMapper 配置，负责分发所有的请求。

（2）HandlerAdapter 选择具体的 Handler 即 controller 处理当前请求。

（3）controller 处理完成之后返回一个与业务相对应的模型和视图交还给前端控制器。

（4）前端控制器根据视图的规则（ViewResolver）解析出具体的视图，然后进行视图渲染。

（5）最后用户看到视图层的结果。

1.4 Spring MVC 之 Hello World！

（1）使用 maven 创建 web-app 工程 helloworld，工程目录结构如图 3 所示。

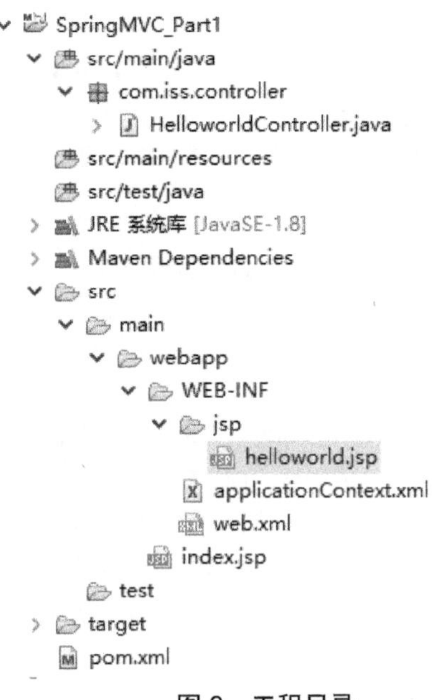

图 3 工程目录

（2）配置 pom.xml 文件，具体内容如下。

```
        // 篇幅有限只截取部分代码
        <dependency>
        <groupId>junit</groupId>
        <artifactId>junit</artifactId>
        <version>4.12</version>
        <scope>test</scope>
    </dependency>
    <dependency>
        <groupId>javax.servlet</groupId>
        <artifactId>javax.servlet-api</artifactId>
        <version>3.0.1</version>
    </dependency>
    <dependency>
        <groupId>org.springframework</groupId>
        <artifactId>spring-core</artifactId>
        <version>4.3.5.RELEASE</version>
    </dependency>
```

Maven 配置 Spring 包如图 4 所示。

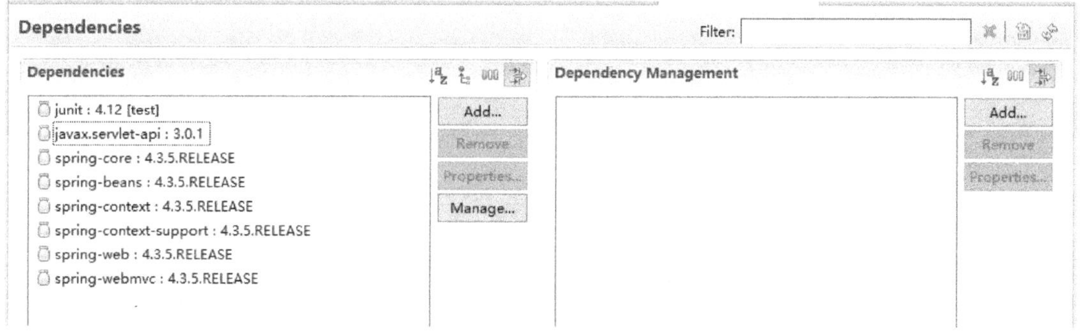

图 4　配置 Spring 包

（3）查看下载的包结构。

可以看到依赖的 jar 包有如下内容，请仔细核对，这里不要引入 asm 包。

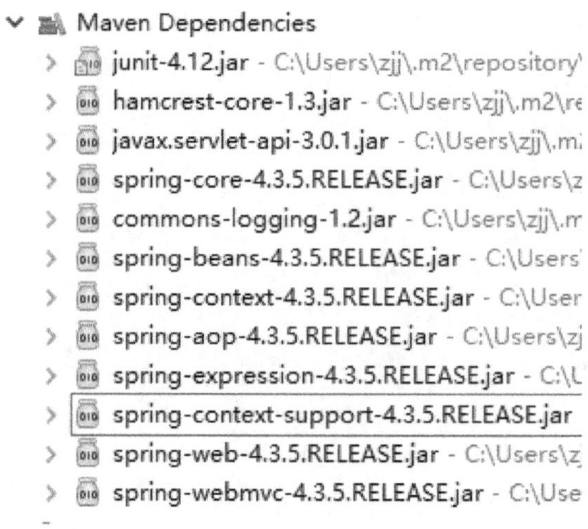

Maven Dependencies
- junit-4.12.jar - C:\Users\zjj\.m2\repository\
- hamcrest-core-1.3.jar - C:\Users\zjj\.m2\re
- javax.servlet-api-3.0.1.jar - C:\Users\zjj\.m
- spring-core-4.3.5.RELEASE.jar - C:\Users\z
- commons-logging-1.2.jar - C:\Users\zjj\.m
- spring-beans-4.3.5.RELEASE.jar - C:\Users
- spring-context-4.3.5.RELEASE.jar - C:\User
- spring-aop-4.3.5.RELEASE.jar - C:\Users\zj
- spring-expression-4.3.5.RELEASE.jar - C:\U
- spring-context-support-4.3.5.RELEASE.jar
- spring-web-4.3.5.RELEASE.jar - C:\Users\z
- spring-webmvc-4.3.5.RELEASE.jar - C:\Use

（4）创建 HelloworldController.java 文件。

```
@Controller
@RequestMapping("/welcome")
public class HelloworldController {
@RequestMapping(method = RequestMethod.GET)
public String printWelcome(HttpSession session) {
        session.setAttribute("info", "welcome to Spring MVC");
                // 简化处理，这里可以使用模型服务等
        return "helloworld";
}
}
```

（5）配置 web.xml 文件。

这里请注意 <url-pattern>/</url-pattern> 的配置。

```
<!DOCTYPE web-app PUBLIC
 "-//Sun Microsystems, Inc.//DTD Web Application 2.3//EN"
 "http://java.sun.com/dtd/web-app_2_3.dtd" >
<web-app>
        <display-name>Archetype Created Web Application</display-name>
        <servlet>
            <servlet-name>springMvc</servlet-name>
            <servlet-class>org.springframework.web.servlet.DispatcherServlet</servlet-class>
          <init-param>
```

```
            <param-name>contextConfigLocation</param-name>
            <param-value>/WEB-INF/applicationContext.xml</param-value>
        </init-param>
    </servlet>
    <servlet-mapping>
        <servlet-name>springMvc</servlet-name>
        <url-pattern>/</url-pattern>
    </servlet-mapping>
</web-app>
```

（6）创建 applicationContext.xml 文件，所在目录为 WEB-INF，具体内容如下。

```
<?xml version="1.0" encoding="UTF-8"?>
<beans xmlns="http://www.springframework.org/schema/beans"
    xmlns:context="http://www.springframework.org/schema/context"
    xmlns:xsi="http://www.w3.org/2001/XMLSchema-instance"
    xsi:schemaLocation="
        http://www.springframework.org/schema/beans
      http://www.springframework.org/schema/beans/spring-beans-4.3.xsd
        http://www.springframework.org/schema/context
       http://www.springframework.org/schema/context/spring-context-4.3.xsd">
    <!—扫描注解的包，包括子集 -->
    <context:component-scan base-package="com.iss.controller" />
    <!—配置视图解析器 -->
    <bean id="viewResolver"
    class="org.springframework.web.servlet.view.InternalResourceViewResolver">
        <property name="prefix" value="/WEB-INF/jsp/" />
        <property name="suffix" value=".jsp"></property>
    </bean>
</beans>
```

（7）创建 index.jsp 文件，具体内容如下。

```
<%@ page language="java" contentType="text/html; charset=UTF-8"
     pageEncoding="UTF-8"%>
<html>
<body>
    <h2>
        <a href="welcome">Print SpringMVC MSG</a>
```

```
    </h2>
</body>
</html>
```

（8）创建 helloworld.jsp 文件，具体内容如下。

```
<%@ page language="java" isELIgnored="false" contentType="text/html; charset=UTF-8"
        pageEncoding="UTF-8"%>
<html>
<head>
<meta http-equiv="Content-Type" content="text/html; charset=UTF-8">
<title>Insert title here</title>
</head>
<body>${sessionScope.info}
</body>
</html>
```

（9）测试方法：将工程部署在 tomcat 上，启动服务器，点击链接，观察浏览器效果。

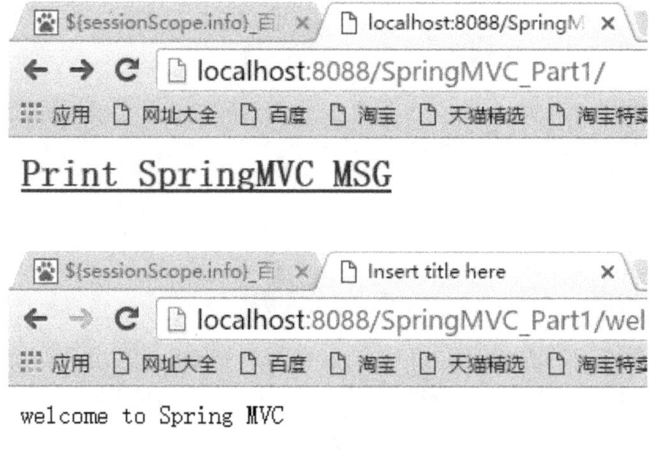

图 5　浏览器效果

小结

（1）Spring MVC 让我们能非常简单地设计出干净的 Web 层和薄薄的 Web 层；
（2）进行更简洁的 Web 层的开发，天生与 Spring 框架集成（如 IoC 容器、AOP 等）；
（3）提供强大的约定大于配置的契约式编程支持；
（4）能简单地进行 Web 层的单元测试；
（5）支持灵活的 URL 到页面控制器的映射；

（6）非常容易与其他视图技术集成，如 Velocity、FreeMarker 等，因为模型数据不放在特定的 API 里，而是放在一个 Model 里（Map 数据结构实现，因此很容易被其他框架使用）；

（7）非常灵活的数据验证、格式化和数据绑定机制，能使用任何对象进行数据绑定，不必实现特定框架的 API；

（8）更加简单的异常处理；

（9）对静态资源的支持；

经典面试题

（1）什么是 Spring MVC 框架？

（2）Spring MVC 有哪些有点？

（3）简述 Spring MVC 处理请求的流程。

（4）解释 DispatcherServlet 前置控制器。

（5）解释 @Controller 注解。

（6）解释 @RequestMapping 注解。

（7）什么是基于注解的容器配置？

（8）怎样开启注解装配？

跟我上机

使用 Spring MVC+Spring JdbcTemplate+BootStrap 框架，采用 XML 配置文件方式完成图 6 所示登录功能。

注意：

①建立 Maven Web 项目，项目结构采用分层方式；

②有登录失败正确和错误提示。

图 6　登录示意图

第 2 章　Spring MVC 配置详解

本章要点（学会后请在方框中打钩）：

- □ 掌握前端控制器的作用
- □ 掌握 Spring MVC 在 web.xml 中的配置
- □ 掌握 Spring MVC 配置文件的配置内容
- □ 掌握 Spring 和 Spring MVC 整合的配置方法
- □ 了解 DispatcherServlet 在上下文的加载顺序

在前面 helloworld 工程中，我们已经展示了基本的 MVC 结构及使用方法。接下来，我们更加深入地学习一下 Spring MVC 中的前端控制器的内容，即 DispatcherServlet。

2.1 DispatcherServlet

DispatcherServlet 是 web 工程中用户访问的入口。是分派调度各个请求的指挥官。是流程控制的核心。

Web 工程中的 web.xml 配置方式具体内容如下。

```
<servlet>
        <servlet-name>springMvc</servlet-name>
        <servlet-class>org.springframework.web.servlet.DispatcherServlet</servlet-class>
        <init-param>
            <param-name>contextConfigLocation</param-name>
            <param-value>classpath:applicationContext.xml</param-value>
        </init-param>
    </servlet>
    <servlet-mapping>
        <servlet-name>springMvc</servlet-name>
        <url-pattern>/</url-pattern>
    </servlet-mapping>
```

专家讲解

（1）<servlet-name>: 其值可以自定义，但是需要注意的是要与下方的 <servlet-mapping> 中的 <servlet-name> 对应，保持一致。

（2）<servlet-class>: 这里配置的就是我们的前端控制器所属的位置，由包名 + 类名组成。

（3）<init-param>：表示初始化的参数，这里如果不配置的话会采用默认的配置方式，为了保证程序正确性，这里我们手动配置了上下文关系配置的名称与配置文件的路径。

（4）<url-pattern>：此处配置的前端控制器需要拦截的请求的地址形式。常见的配置为：*.do，*.htm 等。但是，需要特别注意“/*”表示拦截所有的请求，这时我们将不会在程序中捕获任何请求，因为所有的请求都被前端控制器给拦截下来了。

（5）还有一个常见的配置项：<load-on-startup>1</load-on-startup> 中间的数字表示启动顺序，数字越小表示启动级别越高。注意其最小值为 0，当为负数或者没有指定时，则指示容器在该 servlet 被选择时才加载。

专家讲解

（1）<servlet-mapping>

（2）<servlet-name>springMvc</servlet-name>

（3）<url-pattern>/</url-pattern>

（4） </servlet-mapping>

拦截请求地址的形式讲解内容如下。

（1）拦截 *.do、*.htm，例如：/user/add.do，这是最传统的方式，最简单也最实用。不会导致静态文件（jpg,js,css) 被拦截。

（2）拦截 /，例如：/user/add 可以实现现在很流行的 REST 风格。很多互联网类型的应用很喜欢这种风格的 URL。弊端是会导致静态文件（jpg,js,css) 被拦截后不能正常显示。想实现 REST 风格，事情就会麻烦一些。后面会讲到还算简单的解决办法。

（3）拦截 /*，这是一个错误的方式，请求可以走到 Action 中，但转到 jsp 时再次被拦截，不能访问到 jsp。

2.2 Spring 和 Spring MVC 整合的 web.xml 配置

```
<context-param>
        <param-name>contextConfigLocation</param-name>
        <param-value>
            classpath*:/applicationContext.xml,
            classpath*:/applicationContext-*.xml,
        </param-value>
    </context-param>
    <listener>
<listener-class>org.springframework.web.context.ContextLoaderListener</listener-class>
    </listener>
    <servlet>
        <servlet-name>springServlet</servlet-name>
        <servlet-class>org.springframework.web.servlet.DispatcherServlet</servlet-class>
        <init-param>
            <param-name>contextConfigLocation</param-name>
            <param-value>/WEB-INF/spring-mvc.xml</param-value>
        </init-param>
```

```
            <load-on-startup>1</load-on-startup>
        </servlet>
        <servlet-mapping>
            <servlet-name>springServlet</servlet-name>
            <url-pattern>/</url-pattern>
        </servlet-mapping>
```

专家讲解

（1）contextConfigLocation：表示用于加载 Spring 的 Bean 的配置文件，注意，此名称固定。

（2）listener：监听加载 bean 的实现类，此处也是固定的，默认使用 webapplicationContext。

2.3 spring-mvc.xml 配置

```
    <!-- 自动扫描且只扫描 @Controller -->
    <context:component-scan base-package="com.iss" use-default-filters="false">
    </context:component-scan>
    <!—配置视图解析器 -->
    <bean id="viewResolver"
class="org.springframework.web.servlet.view.InternalResourceViewResolver">
        <property name="prefix" value="/WEB-INF/jsp/" />
        <property name="suffix" value=".jsp"></property>
    </bean>
```

2.4 applicationContext.xml 配置

```
<!-- 使用 annotation 自动注册 bean, 并保证 @Required、@Autowired 的属性被注入 -->
    <context:component-scan base-package="com.iss.dao">
    </context:component-scan>
  <!-- 使用 annotation（注解）开启 AOP -->
    <aop:aspectj-autoproxy/>
    <!-- 使用 annotation 定义事务 -->
    <tx:annotation-driven transaction-manager="transactionManager"/>
</beans>
```

2.5 前端控制器中的上下文加载顺序

（1）项目启动时，容器加载 web.xml。

（2）加载上下文配置即 ContextLoaderListener，此处会将引入的配置文件逐一进行加载。

（3）初始化 MVC 上下文，即加载注解，视图解析器，web 组件等。

（4）由此发现，我们推荐的做法是将上下文配置的位置尽量放置在 web.xml 顶部，防止意外的错误。并且，一旦上下文配置文件加载之后便能够在全局范围内被使用。

另外，从上面的示例配置也可以看出，前端控制器只负责 controller 层的配置解析。其他的配置交给上下文来配置管理。

2.6 Spring MVC 框架控制器结构注解

```
@Controller  ① ß 将 UserController 变成一个 Handler
@RequestMapping("/user")  ② ß 指定控制器映射的 URL
public class UserController {
  @RequestMapping(value = "/register") ③ ß 处理方法对应的 URL，相对于②处的 URL
   public String register( ) {
              return "user/register"; ④ ß 返回逻辑视图名
   }
}
```

2.7 请求映射原理

请求映射原理如图 1 所示。

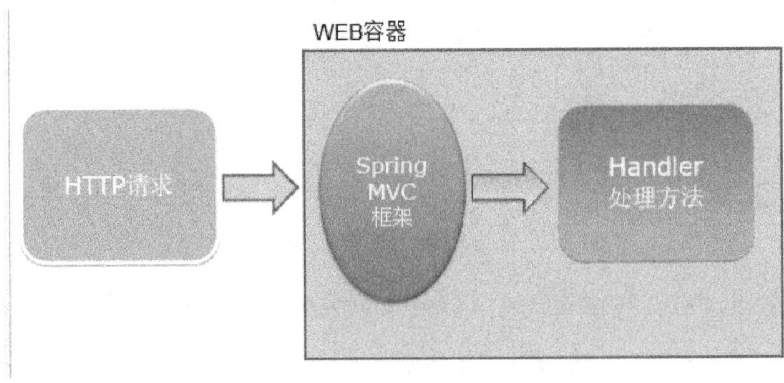

图 1 Spring MVC 请求映射原理

2.8 限定 URL 表达式

1.@RequestMapping 不但支持标准的 URL，还支持 Ant 风格（即 ?、* 和 ** 的字符）的和带 {xxx} 占位符的 URL。以下 URL 都是合法的。

2./user/*/createUser　匹配 /user/aaa/createUser、/user/bbb/createUser 等 URL。

3./user/**/createUser　匹配 /user/createUser、/user/aaa/bbb/createUser 等 URL。

4./user/createUser??　匹配 /user/createUseraa、/user/createUserbb 等 URL。

5./user/{userId}　匹配 user/123、user/abc 等 URL。

6./user/**/{userId}　匹配 user/aaa/bbb/123、user/aaa/456 等 URL。

2.9 通过 URL 限定：绑定 ××× 中的值

```
@RequestMapping("/{userId}")
public ModelAndView showDetail(@PathVariable("userId") String userId){
        ModelAndView mav = new ModelAndView( );
        mav.setViewName("user/showDetail");
        mav.addObject("user", userService.getUserById(userId));
        return mav;
}

@Controller
@RequestMapping("/owners/{ownerId}")
public class RelativePathUriTemplateController {
    @RequestMapping("/pets/{petId}")
    public void findPet(@PathVariable String ownerId,
                                              @PathVariable String petId, Model model) {
      …
    }
```

2.10 通过请求方法限定

示例 1 如下。

```
@RequestMapping(value="/delete")
public String test1(@RequestParam("userId") String userId){
    return "user/test1";
}
à 所有 URL 为 <controllerURI>/delete 的请求由 test1 处理 ( 任何请求方法 )
```

示例 2 如下。

```
@RequestMapping(value="/delete",method=RequestMethod.POST)
public String test1(@RequestParam("userId") String userId){
    return "user/test1";
}
à 所有 URL 为 <controllerURI>/delete 且请求方法为 POST 的请求由 test1 处理
```

小结

（1）Spring MVC 是 Spring 提供的一个强大而灵活的 web 框架。借助于注解，Spring MVC 提供了几乎是 POJO 的开发模式，使得控制器的开发和测试更加简单。这些控制器一般不直接处理请求，而是将其委托给 Spring 上下文中的其他 bean，通过 Spring 的依赖注入功能，这些 bean 被注入到控制器中。

（2）Spring MVC 主要由 DispatcherServlet、处理器映射、处理器 (控制器)、视图解析器、视图组成。

（3）处理器映射：选择使用哪个控制器来处理请求。

（4）视图解析器：选择结果应该如何渲染。

（5）通过以上两点，Spring MVC 保证了如何选择控制处理请求和如何选择视图展现输出之间的松耦合。

经典面试题

（1）为什么要配置 Spring MVC 中的视图解析器 ?

（2）如何向 Spring 容器提供配置元数据 ?

（3）如何在 web 应用里面配置 Spring 和 Spring MVC?

（4）介绍一下配置 Spring MVC 框架 Web 应用的加载顺序是什么。

跟我上机

创建 Maven Web 项目，使用 Spring+Spring MVC+Spring JdbcTemplate 技术完成获取用户信息的功能，效果界面如图 2 所示。

图 2　效果界面

第 3 章　Spring MVC 注解

本章要点（学会后请在方框中打钩）：

- □　掌握 Spring MVC 框架中的各种常用注解
- □　掌握部分注解和配置文件的对比

现在来介绍一下在 spring MVC 中经常使用的注解的含义及使用方式。

3.1 注解配置相对于 XML 配置的优势

（1）它可以充分利用 Java 的反射机制获取类结构信息，这些信息可以有效减少配置的工作。如使用 JPA 注解配置 ORM 映射时，我们就不需要指定 PO 的属性名、类型等信息，如果关系表字段和 PO 属性名、类型都一致，我们甚至无须编写任务属性映射信息——因为这些信息都可以通过 Java 反射机制获取。

（2）注解和 Java 代码位于一个文件中，而 XML 配置采用独立的配置文件，大多数配置信息在程序开发完成后都不会调整，如果配置信息和 Java 代码放在一起，有助于增强程序的内聚性。而采用独立的 XML 配置文件，程序员在编写一个功能时，往往需要在程序文件和配置文件中不停切换，这种思维上的不连贯会降低开发效率。

（3）因此在很多情况下，注解配置比 XML 配置更受欢迎，注解配置有进一步流行的趋势。Spring 2.5 的一大增强就是引入了很多注解类，现在已经可以使用注解配置完成大部分 XML 配置的功能。

3.2 XML 配置 Bean 与 Bean 之间的关系

在使用注解配置之前，先来回顾一下传统做法是如何配置 Bean 并完成 Bean 之间依赖关系的建立。下面有 3 个类，它们分别是 Office、Car 和 Boss，需要在 Spring 容器中配置为 Bean，以下均省略 getter，setter 方法和重写 toString( ) 方法等。

1. office.java

```
public class Office {
        private String officeNo =“001”;
}
```

2. Car.java

```
public class Car {
        private String brand;
        private double price;
}
```

3. Boss.java

```
public class Boss {
    private Car car;
    private Office office;
}
```

我们在 Spring 容器中将 Office 和 Car 声明为 Bean，并注入 Boss Bean 中，下面是使用传统 XML 完成这个工作的配置文件 beans.xml。

```
<?xml version="1.0" encoding="UTF-8" ?>
<beans xmlns="http://www.springframework.org/schema/beans"
    xmlns:xsi="http://www.w3.org/2001/XMLSchema-instance"
    xsi:schemaLocation="http://www.springframework.org/schema/beans
 http://www.springframework.org/schema/beans/spring-beans-2.5.xsd">
    <bean id="boss" class="com.baobaotao.Boss">
        <property name="car" ref="car"/>
        <property name="office" ref="office" />
    </bean>
    <bean id="office" class="com.baobaotao.Office">
        <property name="officeNo" value="002"/>
    </bean>
    <bean id="car" class="com.baobaotao.Car" scope="singleton">
        <property name="brand" value=" 红旗 CA72"/>
        <property name="price" value="2000"/>
    </bean>
</beans>
```

当我们运行以下代码时，控制台将正确打出 boss 的信息。

```
public class AnnoIoCTest {
    public static void main(String[] args) {
        String[] locations = {"beans.xml"};
        ApplicationContext ctx =
            new ClassPathXmlApplicationContext(locations);
        Boss boss = (Boss) ctx.getBean("boss");
        System.out.println(boss);
    }
}
```

这说明 Spring 容器已经正确完成了 Bean 创建和装配的工作。先看一下 Spring MVC

的各种注解的使用。

3.3 Spring MVC 的各种注解使用

3.3.1 @autowired

Spring 2.5 引入了 @Autowired 注解，它可以对类成员变量、方法及构造函数进行标注，完成自动装配的工作。下面是使用 @Autowired 进行成员变量自动注入的代码。

```
public class Boss {
    @Autowired
    private Car car;

    @Autowired
    private Office office;
}
```

Spring 通过一个 BeanPostProcessor 对 @Autowired 进行解析，所以要让 @Autowired 起作用必须事先在 Spring 容器中声明 AutowiredAnnotationBeanPostProcessor Bean。

```
<!-- 该 BeanPostProcessor 将自动起作用，对标注 @Autowired 的 Bean 进行自动注入 -->
    <bean class="org.springframework.beans.factory.annotation.
        AutowiredAnnotationBeanPostProcessor"/>
    <!-- 移除 boss Bean 的属性注入配置的信息 -->
    <bean id="boss" class="com.iss.Boss"/>
    <bean id="office" class="com.iss.Office">
        <property name="officeNo" value="001"/>
    </bean>
    <bean id="car" class="com.iss.Car" scope="singleton">
        <property name="brand" value=" 红旗 CA72"/>
        <property name="price" value="2000"/>
    </bean>
</beans>
```

当然也可以通过 @Autowired 对方法或构造函数进行标注，具体代码内容如下。

```
public class Boss {
        private Car car;
        private Office office;

         @Autowired
        public void setCar(Car car) {
                this.car = car;
        }

        @Autowired
        public void setOffice(Office office) {
                this.office = office;
        }
}
```

这时，@Autowired 将查找被标注的方法的入参类型的 Bean，并调用方法自动注入这些 Bean。而下面的使用方法则会对构造函数进行标注。

```
public class Boss {
        private Car car;
        private Office office;

        @Autowired
        public Boss(Car car ,Office office){
                this.car = car;
                this.office = office ;
        }
    }
```

专家讲解

由于 Boss( ) 构造函数有两个入参，分别是 car 和 office，@Autowired 将分别寻找和它们类型匹配的 Bean，将它们作为 Boss(Car car ,Officeoffice) 的入参来创建 Boss Bean。

3.3.2 @Qualifier

Spring 允许我们通过 @Qualifier 注解指定注入 Bean 的名称，用来消除歧义。

```
@Autowired
public void setOffice(@Qualifier("office") Office office) {
        this.office = office;
}
```

@Qualifier("office") 中的 office 是 Bean 的名称，所以 @Autowired 和 @Qualifier 结合使用时，自动注入的策略就从 byType 转变成 byName 了。@Autowired 可以对成员变量、方法以及构造函数进行注解，而 @Qualifier 的标注对象是成员变量、方法入参、构造函数入参。正是由于注解对象的不同，所以 Spring 不将 @Autowired 和 @Qualifier 统一成一个注解类。

下面是对成员变量进行注解的代码。

```
public class Boss {
        @Autowired
        private Car car;

        @Autowired
        @Qualifier("office")
        private Office office;
}
```

专家讲解

Qualifier 只能和 @Autowired 结合使用，是对 @Autowired 有益的补充。一般来讲，@Qualifier 对方法签名中入参进行注解会降低代码的可读性，而对成员变量注解则相对好一些。

3.3.3 @Resource

Spring 不但支持自己定义的 @Autowired 的注解，还支持几个由 JSR-250 规范定义的注解，它们分别是 @Resource、@PostConstruct 以及 @PreDestroy。

@Resource 的作用相当于 @Autowired，只不过 @Autowired 按 byType 自动注入，而 @Resource 默认按 byName 自动注入罢了。@Resource 有两个属性是比较重要的，分别是 name 和 type，Spring 将 @Resource 注解的 name 属性解析为 Bean 的名字，而 type 属性则解析为 Bean 的类型。所以如果使用 name 属性，则使用 byName 的自动注入策略，而使用 type 属性时则使用 byType 自动注入策略。如果既不指定 name 也不指定 type 属性，这时将通过反射机制使用 byName 自动注入策略。

Resource 注解类位于 Spring 发布包的 common-annotations.jar 类包中，因此在使用之前必须将其加入到项目的类库中。来看一个使用 @Resource 的例子。

```
import javax.annotation.Resource;
public class Boss {
        // 自动注入类型为 Car 的 Bean
        @Resource
        private Car car;

        // 自动注入 bean 名称为 office 的 Bean
        @Resource(name = "office")
        private Office office;
}
```

要让 JSR-250 的注解生效，除了在 Bean 类中标注这些注解外，还需要在 Spring 容器中注册一个负责处理这些注解的 BeanPostProcessor。

```
<bean class="org.springframework.context.annotation.CommonAnnotationBeanPostProcessor"/>
```

CommonAnnotationBeanPostProcessor 实现了 BeanPostProcessor 接口，它负责扫描使用了 JSR-250 注解的 Bean，并对它们进行相应的操作。

3.3.4 @PostConstrust 和 @PreDestroy

Spring 容器中的 Bean 是有生命周期的，Spring 允许在 Bean 在初始化完成后以及 Bean 销毁前执行特定的操作，初始化之后 / 销毁之前方法的指定定义了两个注解类，分别是 @PostConstruct 和 @PreDestroy，这两个注解只能应用于方法上。标注了 @PostConstruct 注解的方法将在类实例化后调用，而标注了 @PreDestroy 的方法将在类销毁之前调用。

```
public class Boss {
        @Resource
        private Car car;

        @Resource(name = "office")
        private Office office;

        @PostConstruct
        public void postConstruct1( ){
            System.out.println("postConstruct1");
        }

        @PreDestroy
```

```
    public void preDestroy1( ){
        System.out.println("preDestroy1");
    }
}
```

通过以下的测试代码，可以看到 Bean 的初始化 / 销毁方法是如何被执行的。

```
public class AnnoIoCTest {
    public static void main(String[] args) {
        String[] locations = {"beans.xml"};
        ClassPathXmlApplicationContext ctx =
            new ClassPathXmlApplicationContext(locations);
        Boss boss = (Boss) ctx.getBean("boss");
        System.out.println(boss);
        ctx.destroy( );// 关闭 Spring 容器，以触发 Bean 销毁方法的执行
    }
}
```

这时，将看到标注了 @PostConstruct 的 postConstruct1( ) 方法将在 Spring 容器启动时，创建 Boss Bean 的时候被触发执行，而标注了 @PreDestroy 注解的 preDestroy1( ) 方法将在 Spring 容器关闭前销毁 Boss Bean 的时候被触发执行。

前面所介绍的 AutowiredAnnotationBeanPostProcessor 和 CommonAnnotationBeanPostProcessor 就是处理这些注解元数据的处理器。但是直接在 Spring 配置文件中定义这些 Bean 显得比较笨拙。Spring 为我们提供了一种方便的注册这些 BeanPostProcessor 的方式，这就是 <context:annotation-config/>。请看下面的配置。

```
// 引入上下文命名空间
<context:annotation-config/>
    <bean id="boss" class="com.iss.Boss"/>
    <bean id="office" class="com.iss.Office">
        <property name="officeNo" value="001"/>
    </bean>
    <bean id="car" class="com.iss.Car" scope="singleton">
        <property name="brand" value=" 红旗 CA72"/>
        <property name="price" value="2000"/>
    </bean>
</beans>
```

<context:annotationconfig/> 将隐式地向 Spring 容器注册 AutowiredAnnotationBeanPostProcessor、CommonAnnotationBeanPostProcessor、PersistenceAnnotationBeanPostProcessor 以

及 equiredAnnotationBeanPostProcessor 这 4 个 BeanPostProcessor。

在配置文件中使用 context 命名空间之前，必须在 <beans> 元素中声明 context 命名空间。

3.3.5 @Component

虽然我们可以通过 @Autowired 或 @Resource 在 Bean 类中使用自动注入功能，但是 Bean 还是在 XML 文件中通过 <bean> 进行定义 —— 也就是说，在 XML 配置文件中定义 Bean，通过 @Autowired 或 @Resource 为 Bean 的成员变量、方法入参或构造函数入参提供自动注入的功能。能否也通过注解定义 Bean，从 XML 配置文件中完全移除 Bean 定义的配置呢？答案是肯定的，我们通过 Spring 提供的 @Component 注解就可以达到这个目标了。

为什么 @Repository 只能标注在 DAO 类上呢？这是因为该注解的作用不只是将类识别为 Bean，同时它还能将所标注的类中抛出的数据访问异常封装为 Spring 的数据访问异常类型。 Spring 本身提供了一个丰富的并且是与具体的数据访问技术无关的数据访问异常结构，用于封装不同的持久层框架抛出的异常，使得异常独立于底层的框架。

Spring 在 @Repository 的基础上增加了功能类似的额外三个注解：@Component、@Service、@Controller，它们分别用于软件系统的不同层次。

- @Component 是一个泛化的概念，仅仅表示一个组件 (Bean) ，可以作用在任何层次。
- @Service 通常作用在业务层，但是目前该功能与 @Component 相同。
- @Controller 通常作用在控制层，但是目前该功能与 @Component 相同。

通过在类上使用 @Repository、@Component、@Service 和 @Constroller 注解，Spring 会自动创建相应的 BeanDefinition 对象，并注册到 ApplicationContext 中。这些类就成了 Spring 受管组件。这三个注解除了作用于不同软件层次的类，其使用方式与 @Repository 是完全相同的。

这样，我们就可以在 Boss 类中通过 @Autowired 注入前面定义的 Car 和 Office Bean 了。

```
@Component("boss")
public class Boss {
    @Autowired
    private Car car;

    @Autowired
    private Office office;
}
```

在使用 @Component 注解后，Spring 容器必须启用类扫描机制以启用注解驱动 Bean 定义和注解驱动 Bean 自动注入的策略。Spring 对 context 命名空间进行了扩展，请看下面的配置。

```
<context:component-scan base-package="com.iss"/>
```

这里，所有通过 <bean> 元素定义 Bean 的配置内容已经被移除，仅需要添加一行 <context:component-scan/> 配置就解决所有问题了，Spring XML 配置文件得到了极致的简化。<context:component-scan/> 的 base-package 属性指定了需要扫描的类包，类包及其递归子包中所有的类都会被处理。

3.3.6 @Scope

默认情况下通过 @Component 定义的 Bean 都是 singleton 的，如果需要使用其他作用范围的 Bean，可以通过 @Scope 注解来达到目标，如以下代码所示。

```
@Scope("prototype")
@Component("boss")
public class Boss {
    …
}
```

这样，当从 Spring 容器中获取 boss Bean 时，每次返回的都是新的实例了。

3.3.7 @Controller

@Controller 对应表现层的 Bean，也就是 *Action，*Controller 如下所示。

```
@Controller
@Scope("prototype")
public class UserController extends BaseController <User>{
 ……
}
```

3.3.8 @Service

@Service 对应的是业务层 Bean，如下所示。

```
@Service("userService")
public class UserServiceImpl implements UserService {
………
}
```

@Service("userService") 注解是告诉 Spring，当 Spring 要创建 UserServiceImpl 的实例时，bean 的名字必须叫做"userService"，这样当 Action 需要使用 UserServiceImpl 的实例时，就可以由 Spring 创建好的"userService"，然后注入给 Action：在 Action 只需要声明一个名字叫"userService"的变量来接收由 Spring 注入的"userService"即可，具体代码如下。

```
// 注入 userService
@Resource(name = "userService")
private UserService userService;
```

注意：在 Action（controller 层）声明的“userService”变量的类型必须是“UserServiceImpl”或者是其父类“UserService”，否则会因为类型不一致而无法注入。

3.3.9 @Repository

@Repository 对应数据访问层 Bean，如下所示。

```
@Repository(value="userDao")
public class UserDaoImpl extends BaseDaoImpl<User> {
………
}
```

@Repository(value="userDao") 注解是告诉 Spring，让 Spring 创建一个名字叫“userDao”的 UserDaoImpl 实例。当 Service 需要使用 Spring 创建的名字叫“userDao”的 UserDaoImpl 实例时，就可以使用 @Resource(name = "userDao") 注解告诉 Spring，Spring 把创建好的 userDao 注入给 Service 即可。

```
// 注入 userDao，从数据库中根据用户 Id 取出指定用户时需要用到
@Resource(name = "userDao")
private BaseDao<User> userDao;
```

3.3.10 @RequestMapping

@RequestMapping 是一个用来处理请求地址映射的注解，可用于类或方法上。用于类上，表示类中的所有响应请求的方法都以该地址作为父路径。

RequestMapping 注解有六个属性，下面我们把它分成两类进行说明。

1.value 与 method

value：指定请求的实际地址，指定的地址可以是 URI Template 模式。

value 的 uri 值为以下两类。

（1）可以指定为普通的具体值。

```
@RequestMapping("/appointments")
public class AppointmentsController {
...
}
```

（2）可以指定为含有某变量的一类值。

```
@RequestMapping(value="/owners/{ownerId}", method=RequestMethod.GET)
public String findOwner(@PathVariable String ownerId, Model model) {
  Owner owner = ownerService.findOwner(ownerId);
  model.addAttribute("owner", owner);
  return "displayOwner";
}
```

method：指定请求的 method 类型，GET、POST、PUT、DELETE 等。

```
@RequestMapping(method = RequestMethod.GET)
  public Map<String, Appointment> get( ) {
       return appointmentBook.getAppointmentsForToday( );
  }
```

2.consumes 与 produces

consumes：指定处理请求的提交内容类型（Content-Type），例如 application/json,text/html。

```
@Controller
@RequestMapping(value = "/pets", method = RequestMethod.POST, consumes="application/json")
public void addPet(@RequestBody Pet pet, Model model) {
      // implementation omitted
}
```

produces: 指定返回的内容类型，仅当 request 请求头中的 (Accept) 类型中包含该指定类型才返回；同时暗示了返回的内容类型为 application/json。

```
@Controller
@RequestMapping(value = "/pets/{petId}", method = RequestMethod.GET, produces="application/json")
@ResponseBody
public Pet getPet(@PathVariable String petId, Model model) {
      // implementation omitted
}
```

3.3.11 @RequestParam

（1）常用来处理简单类型的绑定，通过 Request.getParameter( ) 获取的 String 可直接转换为简单类型的情况（ String--> 简单类型的转换操作由 ConversionService 配置的转换器来完成）；因为使用 request.getParameter( ) 方式获取参数，所以可以处理 get 方式中 queryString 的值，也可以处理 post 方式中 body data 的值。

（2）用来处理 Content-Type: 为 application/x-www-form-urlencoded 编码的内容，提交方式 GET、POST。

（3）该注解有两个属性：value、required；value 用来指定要传入值的 id 名称，required 用来指示参数是否必须绑定。

示例代码 1：

```
@RequestMapping("testRequestParam")
public String filesUpload(@RequestParam(value="aa", required=true) String inputStr,
 HttpServletRequest request)
```

上面的 required=true 或者 false 表示，当传入参数中必须传入或者可以不传并且将 aa 赋值为 null，【一定是 null】。

所以请特别注意：required=false 只能用来改变是否传入报错，而不能改变传入值的参数类型，即如果 aa 定义为 int 类型，并且前台没有传入的情况下将会引起数据类型转化的错误。解决办法是使用封装的 Integer 来代替声明的 int。

3.3.12 @RequestBody

（1）用来接收一个 String 化的 json，前端常见的做法是使用 JSON.stringify(json) 这个方法来转化，并且使用 post 发出。

（2）要设置 contentType，contentType:"application/json，明确告诉服务器发送的内容是 json，而默认的 contentType 是 application/x-www-form-urlencoded; charset=UTF-8。

特别注意：前台传入的 json 中的 key 在实体中必须要存在，不然就会报错。

示例代码如下。

```
@RequestMapping("/repairs/saveDispatches")
    public void saveDispatches(@RequestBody DispatchesDTO dispatchesDTO,
            HttpServletResponse response) throws IOException {
        dispatchesService.saveDispatches(dispatchesDTO);
        success(response);
}
```

还需要注意的是，使用该配置，必须加入如下配置。

```
<!--Spring3.1 开始的注解 HandlerAdapter -->
<bean class="org.springframework.web.servlet.mvc.method.annotation.RequestMap-
pingHandlerAdapter">
    <property name="messageConverters">
        <list>
```

```
            <bean class="org.springframework.http.converter.json.MappingJacksonHttpMessageConverter"></bean>
        </list>
    </property>
</bean>
```

或者使用如下配置。

```
<mvc:annotation-driven/>
```

最后还要注意以下两点。

（1）若前台发送请求时未指定 contentType 为 json/application 或 Spring 中未配置 messageConverter，那么会报错误 415，也就是类型不支持。

（2）由于 Spring 中使用的 json 转换用到的是 jackson，所以需要引入 jackson 包。

3.3.13 @ResponseBody

@ResponseBody 表示该方法的返回结果直接写入 HTTP response body 中。

一般在异步获取数据时使用，在使用 @RequestMapping 后，返回值通常解析为跳转路径，加上 @Responsebody 后返回结果不会被解析为跳转路径，而是直接写入 HTTP response body 中。比如异步获取 json 数据，加上 @ResponseBody 后，会直接返回 json 数据。

示例代码如下。

```
@RequestMapping(value = "person/profile/{id}/{name}/{status}")
@ResponseBody
    public Person porfile(@PathVariable int id,@PathVariable String name,@PathVariable boolean status) {
        return new Person(id, name, status);
    }
```

3.3.14 @PathVariable

当使用 @RequestMapping URI template 样式映射时，即 someUrl/{paramId}，这时的 paramId 可通过 @Pathvariable 注解将它传过来的值绑定到方法的参数上。

示例代码如下。

```
@Controller
@RequestMapping("/owners/{ownerId}")
publicclass RelativePathUriTemplateController {
 @RequestMapping("/pets/{petId}")
```

```
    publicvoid findPet(@PathVariable String ownerId, @PathVariable String petId, Model model) {
        // implementation omitted
    }
}
```

假如不是一一对应的关系，可以参照如下的书写方式。

```
@RequestMapping(value = "/person/profile/{id}", method = RequestMethod.GET)
public @ResponseBody
Person porfile(@PathVariable("id") int uid) {
        return new Person(uid, name, status);
}
```

3.3.15 @ModelAttribute

（1）前台页面提交请求并附带 user 相关参数，会把参数赋值给这个 user 对象，同时这个 user 对象也可以带回到前台页面，页面可以用 EL 表达式来读取。此方法会先从 model 去获取 key 为 "user" 的对象，如果获取不到会通过反射实例化一个 User 对象，再从 request 里面拿值 set 到这个对象，然后把这个 User 对象添加到 model(其中 key 为 "user")。

示例代码 1 如下。

```
@RequestMapping(value = "/helloWorld")
            public String helloWorld(@ModelAttribute("user") User user) {
                user.setUserName("jizhou");
                return "helloWorld";
    }
```

（2）注解到 controller 层中的方法上：该 Controller 的所有方法在调用前，先执行此 @ModelAttribute 方法。

示例代码 2 如下。

```
@Controller
@RequestMapping(value="test")
public class PassportController {
    @ModelAttribute
     public void preRun( ) {
          System.out.println("Test Pre-Run");
     }
     @RequestMapping(value="login", method=RequestMethod.POST)
```

```
        public ModelAndView login(@ModelAttribute @Valid Account account, BindingResult result)
        }
}
```

需要特别注意的是，如果出现在 controller 层中，既有方法层面的注解，也有参数的注解。那么参数注解会覆盖方法注解的内容。

特殊场景：

```
@ModelAttribute("user1")
public User addUser(User user) {
        return new User(520,"I love U");
}
```

假设此方法是写在 UserController 内的，那么执行 UserController 内带有 @RequestMapping 的方法之前，都会先执行此 addUser 方法，并且执行 addUser 过程中会添加两个对象到 model, 先添加 key 为 "user" 的对象 (由 addUser 方法的 User user 引起的), 再添加 key 为 "user1" 的对象 (由注解 @ModelAttribute("user1") 引起的)。

（3）从 form 表单或者 URL 参数中获取对象（实际应用时即使没有注解也能获得该对象）。

特别注意的是，此时 User 类一定要有没有参数的构造函数。

```
@Controller
        public class HelloWorldController {
                @RequestMapping(value = "/helloWorld")
                public String helloWorld(@ModelAttribute User user) {
                        return "helloWorld";
        }
    }
```

3.3.16 @SessionAttributes

在默认情况下，ModelMap 中的属性作用域是 request 级别的，也就是说，当本次请求结束后，ModelMap 中的属性将销毁。如果希望在多个请求中共享 ModelMap 中的属性，必须将其属性转存到 Session 中，这样 ModelMap 的属性才可以被跨请求访问。

Spring 允许我们有选择地指定 ModelMap 中的哪些属性需要转存到 Session 中，以便下一个请求属对应的 ModelMap 的属性列表中还能访问到这些属性。这一功能是通过类定义处标注 @SessionAttributes 注解来实现的。

示例代码如下。

```
@Controller
@RequestMapping("/bbtForum.do")
@SessionAttributes("currUser") // ①将 ModelMap 中属性名为 currUser 的属性
// 放到 Session 属性列表中，以便这个属性可以跨请求访问
public class BbtForumController {
…
    @RequestMapping(params = "method=listBoardTopic")
    public String listBoardTopic(@RequestParam("id")int topicId, User user,
ModelMap model) {
        bbtForumService.getBoardTopics(topicId);
        System.out.println("topicId:" + topicId);
        System.out.println("user:" + user);
        model.addAttribute("currUser",user); // ②向 ModelMap 中添加一个属性
        return "listTopic";
    }
}
```

我们在 ② 处添加了一个 ModelMap 属性，其属性名为 currUser，而 ① 处通过 @SessionAttributes 注解将 ModelMap 中名为 currUser 的属性放置到 Session 中，所以我们不但可以在 listBoardTopic( ) 请求所对应的 JSP 视图页面中通过 request.getAttribute(“currUser”)和 session.getAttribute(“currUser”) 获取 user 对象，还可以在下一个请求所对应的 JSP 视图页面中通过 session.getAttribute(“currUser”) 或 ModelMap#get(“currUser”) 访问到这个属性。

特别用法：

我们可以在需要访问 Session 属性的 controller 上加上 @SessionAttributes，然后在 action 需要的 User 参数上加上 @ModelAttribute，并保证两者的属性名称一致。Spring MVC 就会自动将 @SessionAttributes 定义的属性注入到 ModelMap 对象，在 setup action 的参数列表时，去 ModelMap 中取到这样的对象，再添加到参数列表。只要我们不去调用 SessionStatus 的 setComplete( ) 方法，这个对象就会一直保留在 Session 中，从而实现 Session 信息的共享。

最后，特别注意的是，@SessionAttributes 只能声明在类上，而不能声明在方法上。

3.3.17 @Required

@Required 负责检查一个 bean 在初始化时其声明的 set 方法是否被执行，当某个被标注了 @Required 的 Setter 方法没有被调用，则 Spring 在解析的时候会抛出异常，以提醒开发者对相应属性进行设置。 @Required 注解只能标注在 Setter 方法之上。因为依赖注入的本质是检查 Setter 方法是否被调用了，而不是真的去检查属性是否赋值了以及赋了什么样的值，不会测试属性是否非空。如果将该注解标注在非 setXxxx( ) 类型的方法则被忽略。

示例代码如下。

```
@Required
        public void setProduct(Product product) {
                this.product = product;
        }
```

如果没有设置属性的话，抛出 BeanInitializationException 异常。

3.3.18 @InitBinder

在实际操作中经常会碰到表单中的日期字符串和 Javabean 中的日期类型的属性自动转换，而 Spring MVC 默认不支持这个格式的转换，所以必须要手动配置，自定义数据类型的绑定才能实现这个功能。解决的办法就是使用 Spring MVC 提供的 @InitBinder 标签。

示例代码：

```
@InitBinder
public void initBinder(WebDataBinder binder) {
                SimpleDateFormat dateFormat = new SimpleDateFormat("yyyy-MM-dd");
                dateFormat.setLenient(false);
                binder.registerCustomEditor(Date.class, new CustomDateEditor(dateFormat, true));
}
```

注意被标注 @InitBinder 注解的方法必须拥有一个 WebDataBinder 类型的入参，以便 Spring MVC 框架将注册属性编辑器的 WebDataBinder 对象传递进来。

3.3.19 @Value

作用：在代码中注入 *.properties 文件中的数据。

步骤 1 示例代码如下。

```
<bean id="configProperties"
 class="org.springframework.beans.factory.config.PropertiesFactoryBean">
                <property name="locations">
                        <list>
                                <value>classpath:/config/*.properties</value>
                        </list>
                </property>
    </bean>
<bean id="propertyConfigurer"
class="org.springframework.beans.factory.config.PreferencesPlaceholderConfigurer">
                <property name="properties" ref="configProperties" />
</bean>
```

步骤 2 示例代码如下。

```
userPageSize=5 // 此为 properties 文件中的数据
```

步骤 3 示例代码如下。

在 Controller 中使用注解获得配置项内容。

```
@Value("#{configProperties['userPageSize']}")
private String userPageSize;
```

后面的代码就可以使用 userPageSize 这个私有变量了，这个字符串的值就是我们配置文件中配置的 5。

3.3.20 @RestController

为了方便 Rest 开发，Spring 4.0 一个重要的新的改进是 @RestController 注解，它继承自 @Controller 注解。4.0 之前的版本，Spring MVC 的组件都使用 @Controller 来标识当前类是一个控制器 servlet。在 Controller 上标注了 @RestController，这样相当于 Controller 的所有方法都标注了 @ResponseBody，这样就不需要在每个 @RequestMapping 方法上加 @ResponseBody 了。

示例代码如下。

```
@RestController
public class UserController {
        private UserService userService;
        @Autowired
        public UserController(UserService userService) {
                this.userService = userService;
        }
        @RequestMapping("/test")
         public User view( ) {
                User user = new User( );
                user.setId(1L);
                user.setName("haha");
                return user;
        }

        @RequestMapping("/test2")
        public String view2( ) {
                return "{\"id\" : 1}";
        }
}
```

添加了一个 AsyncRestTemplate，支持 REST 客户端的异步无阻塞支持。

示例代码如下。

```
@RestController
public class UserController {
    private UserService userService;
    @Autowired
    public UserController(UserService userService) {
        this.userService = userService;
    }
    @RequestMapping("/api")
      public Callable<User> api( ) {
        System.out.println("=====hello");
        return new Callable<User>( ) {
            @Override
            public User call( ) throws Exception {
                Thread.sleep(10L * 1000); // 暂停两秒
                User user = new User( );
                user.setId(1L);
                user.setName("haha");
                return user;
            }
        };
    }
}
```

测试程序如下。

```
public static void main(String[] args) {
    AsyncRestTemplate template = new AsyncRestTemplate( );
    // 调用完后立即返回（没有阻塞）
    ListenableFuture<ResponseEntity<User>> future =
template.getForEntity("http://localhost:9080/spring4/api", User.class);
    // 设置异步回调
    future.addCallback(new ListenableFutureCallback<ResponseEntity<User>>( ) {
        @Override
        public void onSuccess(ResponseEntity<User> result) {
            System.out.println("======client get result : " +     result.getBody( ));
        }
```

```
            @Override
            public void onFailure(Throwable t) {
                System.out.println("======client failure : " + t);
            }
        });
        System.out.println("==no wait");
    }
```

此处使用 Future 来完成非阻塞，这样的话我们也需要给它一个回调接口来拿结果；Future 和 Callable 是一对，一个消费结果，一个产生结果。调用完模板后会立即返回，不会阻塞；有结果时会调用其回调。

AsyncRestTemplate 默认使用 SimpleClientHttpRequestFactory，即通过 java.net.HttpURLConnection 实现；另外我们也可以使用 apache 的 http components；使用 template.setAsyncRequestFactory(new HttpComponentsAsyncClientHttpRequestFactory( )); 设置即可。

小结

Spring 常用注解汇总。

（1）使用注解之前要开启自动扫描功能，其中 base-package 为需要扫描的包（含子包）。

（2）<context:component-scan base-package="com.iss"/>。

（3）@Configuration 把一个类作为一个 IoC 容器，它的某个方法头上如果注册了 @Bean，就会作为这个 Spring 容器中的 Bean。

（4）@Scope 注解作用域。

（5）@Lazy(true) 表示延迟初始化。

（6）@Service 用于标注业务层组件。

（7）@Controller 用于标注控制层组件（如 struts 中的 action）。

（8）@Repository 用于标注数据访问组件，即 DAO 组件。

（9）@Component 泛指组件，当组件不好归类的时候，我们可以使用这个注解进行标注。

（10）@Scope 用于指定 scope 作用域的（用在类上）。

（11）@PostConstruct 用于指定初始化方法（用在方法上）。

（12）@PreDestory 用于指定销毁方法（用在方法上）。

（13）@DependsOn：定义 Bean 初始化及销毁时的顺序。

（14）@Primary：自动装配时当出现多个 Bean 候选者时，被注解为 @Primary 的 Bean 将作为首选者，否则将抛出异常。

（15）@Autowired 默认按类型装配，如果我们想使用按名称装配，可以结合 @Qualifier 注解一起使用。

（16）@Autowired @Qualifier("personDaoBean") 存在多个实例配合使用。
（17）@Resource 默认按名称装配，当找不到与名称匹配的 bean 才会按类型装配。
（18）@PostConstruct 初始化注解。
（19）@PreDestroy 摧毁注解，默认单例启动就加载。
（20）@Async 异步方法调用。

经典面试题

（1）如何开启基于基于注解的自动写入？
（2）请举例说明 @Required 注解。
（3）请举例说明 @Scope 注解。
（4）请举例说明 @Qualifier 注解。
（5）Spring MVC 如何生成 JSON 数据？注解如何配置？

跟我上机

使用 Spring+Spring MVC+Spring JDBC 全注解配置完成“航班信息管理工具”功能，界面功能如下。

①添加航班信息。

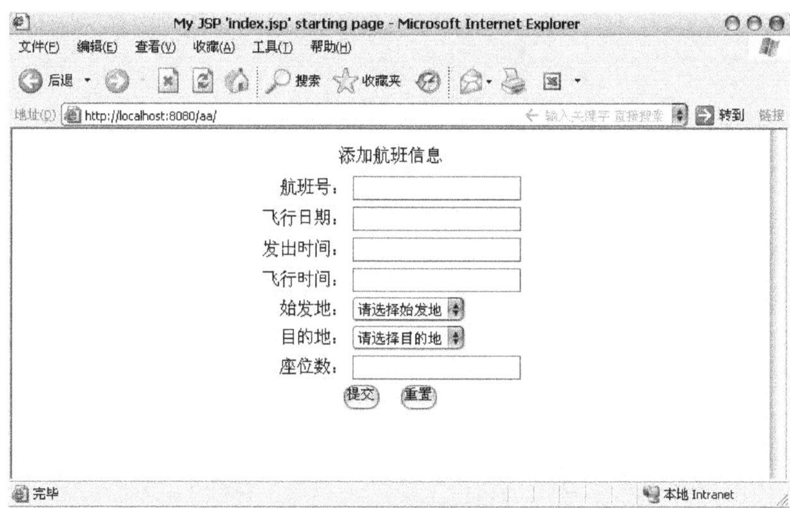

图 1　添加航班信息

②以始发地和目的地名称（须从下拉框选择，不可手动输入）为条件查询航班信息。

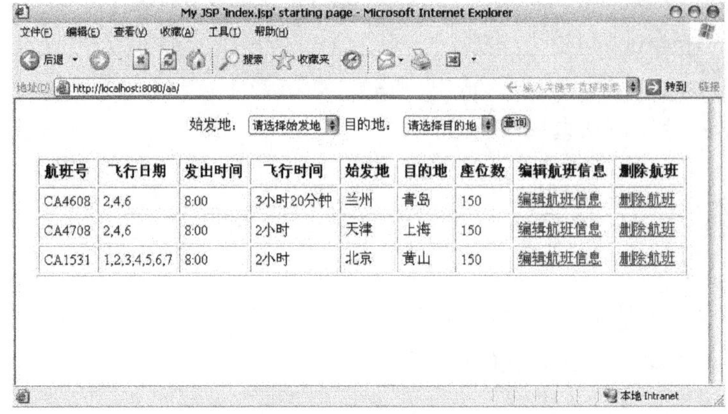

图 2 查询航班信息

③编辑航班信息。

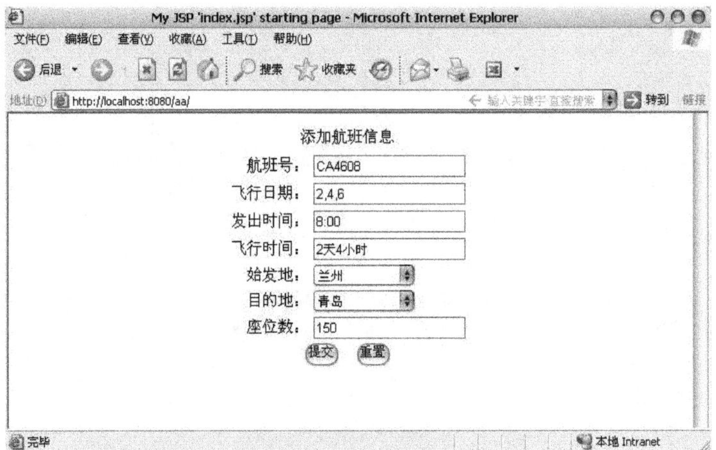

图 3 编辑航班信息

④删除航班信息。

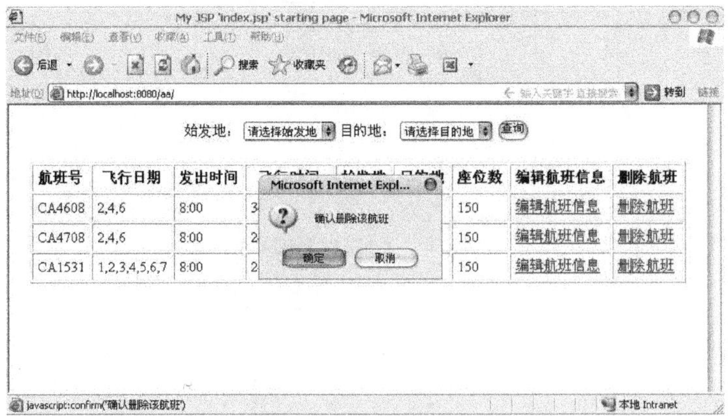

图 4 删除航班信息

参考数据表如表 1、表 2 所示。

表 1　flightinfo 表结构

字段名	类型	约束	描述
flightid	NUMBER(4)	主键	航班 id 序列增长
flightnum	VARCHAR2(10)	非空	航班号
flydate	VARCHAR2(10)	非空	飞行日期（例：每星期 2,4,6）
starttime	VARCHAR2(10)	非空	发出时间（例：8:00）
flytime	VARCHAR2(10)	非空	飞行时间（例：2 天）
startcity	NUMBER（4）	外键	始发地
endcity	NUMBER（4）	外键	目的地
seatnum	NUMBER（4）	非空	座位总数

表 2　cityinfo 表结构

字段名	类型	约束	描述
cityid	NUMBER(2)	主键	城市 id
cityname	VARCHAR2(10)	非空	城市名称

第 4 章　Spring MVC 拦截器

本章要点(学会后请在方框中打钩):

- □ 掌握如何配置 Spring MVC 拦截器
- □ 掌握如何配置多个拦截器
- □ 掌握拦截器和过滤器的区别
- □ 掌握项目开发过程中乱码问题的解决方法

本章来演示 Spring MVC 中使用的拦截器的配置与使用方法。

4.1 配置 Spring MVC 拦截器

（1）使用 maven 创建 web-app 工程 interceptor 工程

（2）配置 web.xml 文件，具体内容如下。

```
<?xml version="1.0" encoding="UTF-8"?>
<web-app xmlns:xsi="http://www.w3.org/2001/XMLSchema-instance"
xmlns="http://java.sun.com/xml/ns/javaee"
xsi:schemaLocation="http://java.sun.com/xml/ns/javaee
http://java.sun.com/xml/ns/javaee/web-app_2_5.xsd" id="WebApp_ID" version="2.5">
    <servlet>
        <servlet-name>springMvc</servlet-name>
        <servlet-class>org.springframework.web.servlet.DispatcherServlet</servlet-class>
        <init-param>
            <param-name>contextConfigLocation</param-name>
            <param-value>/WEB-INF/spring-mvc.xml</param-value>
        </init-param>
    </servlet>
    <servlet-mapping>
        <servlet-name>springMvc</servlet-name>
        <url-pattern>/</url-pattern>
    </servlet-mapping>
</web-app>
```

（3）配置 spring-mvc.xml 文件，具体内容如下。

```
    <context:component-scan base-package="interceptor" />
    <!—配置视图处理转换器 -->
    <bean id="viewResolver"
 class="org.springframework.web.servlet.view.InternalResourceViewResolver">
        <property name="prefix" value="/WEB-INF/jsp/"></property>
        <property name="suffix" value=".jsp"></property>
    </bean>
<!—配置拦截器 -->
    <mvc:interceptors>
        <mvc:interceptor>
```

```
                <mvc:mapping path="/helloworld"/>
            <bean class="interceptor.HelloWorldInterceptor"></bean>
        </mvc:interceptor>
    </mvc:interceptors>
```

（4）创建 HelloWorldController.java 文件，具体内容如下。

```
package interceptor;
@Controller
@RequestMapping(value="helloworld")
public class HelloWorldController {
    @RequestMapping(method=RequestMethod.POST)
    public ModelAndView sayHello(String fname,String lname){
        System.out.println("hello");
        ModelAndView mv = new ModelAndView( );
        System.out.println("fname:"+fname);
        System.out.println("lname:"+lname);
        mv.setViewName("/hello");
        return mv;
    }
}
```

（5）创建 index.jsp，hello.jsp，具体内容如下。

```
<%@ page language="java" contentType="text/html; charset=UTF-8" pageEncoding="UTF-8"%>
<html>
<body>
    <form action="helloworld" method="post">
    <p>First name: <input type="text" name="fname" /></p>
    <p>Last name: <input type="text" name="lname" /></p>
    <input type="submit" value="Submit" />
</form>
</body>
</html>
<%@ page language="java" contentType="text/html; charset=UTF-8"
pageEncoding="UTF-8"%>
<html>
<body>
```

```
helloworld
</body>
</html>
```

（6）创建 HelloWorldInterceptor.java 文件，具体内容如下。

```
public class  HelloWorldInterceptor implements HandlerInterceptor {
        /**
          preHandle 方法是进行处理器拦截用的，顾名思义，该方法将在 Controller 处理之前进行调用， Spring MVC 中的 Interceptor 拦截器是链式的，可以同时存在多个 Interceptor，然后 Spring MVC 会根据声明的前后顺序一个接一个的执行，而且所有的 Interceptor 中的 preHandle 方法都会在 Controller 方法调用之前调用。Spring MVC 的这种 Interceptor 链式结构也是可以进行中断的，这种中断方式是令 preHandle 的返回值为 false，当 preHandle 的返回值为 false 的时候整个请求就结束了。
         */
        @Override
        public boolean preHandle(HttpServletRequest request,
                    HttpServletResponse response, Object handler) throws Exception {
            System.out.println("preHandle");
            return true;
        }

        /**
        这个方法只会在当前这个 Interceptor 的 preHandle 方法返回值为 true 的时候才会执行。postHandle 是进行处理器拦截用的，它的执行时间是在处理器进行处理之后，也就是在 Controller 的方法调用之后执行，但是它会在 DispatcherServlet 进行视图的渲染之前执行，也就是说在这个方法中你可以对 ModelAndView 进行操作。这个方法的链式结构跟正常访问的方向是相反的，也就是说先声明的 Interceptor 拦截器该方法反而会后调用，这跟 Struts2 里面的拦截器的执行过程有点像，只是 Struts2 里面的 intercept 方法中要手动的调用 ActionInvocation 的 invoke 方法， Struts2 中调用 ActionInvocation 的 invoke 方法就是调用下一个 Interceptor 或者是调用 action，然后要在 Interceptor 之前调用的内容都写在调用 invoke 之前，要在 Interceptor 之后调用的内容都写在调用 invoke 方法之后。
           */
          @Override
          public void postHandle(HttpServletRequest request,
                    HttpServletResponse response, Object handler,
                    ModelAndView modelAndView) throws Exception {
```

```
                // TODO Auto-generated method stub
                  System.out.println("postHandle");
        }

    /**
     该方法也是需要当前对应的 Interceptor 的 preHandle 方法的返回值为 true 时才会执行。该方法将在整个请求完成之后，也就是 DispatcherServlet 渲染了视图执行，这个方法的主要作用是用于清理资源的，当然这个方法也只能在当前这个 Interceptor 的 preHandle 方法的返回值为 true 时才会执行。
     */
    @Override
    public void afterCompletion(HttpServletRequest request,
            HttpServletResponse response, Object handler, Exception ex)
    throws Exception {
        // TODO Auto-generated method stub
          System.out.println("afterCompletion");
    }
}
```

专家讲解

```
(1) <!-- 对静态资源文件的访问  方案一（二选一）-->
(2)       <mvc:default-servlet-handler/>
(3) <!-- 对静态资源文件的访问  方案二（二选一）-->
(4)       <mvc:resources mapping="/images/**" location="/images/"
  cache-period="31556926"/>
(5)       <mvc:resources mapping="/js/**" location="/js/" cache-period="31556926"/>
(6)       <mvc:resources mapping="/css/**" location="/css/" cache-period="31556926"/>
```

4.2 Spring MVC 多个拦截器

上文我们介绍了 Spring 中过滤器的基本用法，现在我们来介绍多个拦截器的执行情况，另外一种拦截器的实现方式，以及拦截器与 java 过滤器的区别。

（1）在上文中，我们创建了 HelloWorldInterceptor.java 文件，在此基础之上，我们再创建一个 HelloWorldInterceptor2.java 文件，文件与上面的内容基本保持一致，只需在输出语句中标识出当前的拦截器名称即可，如下。

```
System.out.println("preHandle2");
```

（2）在上文的配置文件中，我们演示了如何针对单一的 controller 进行拦截，现在，我们来介绍如何在全局范围内使用拦截器。具体配置如下。

```
<context:component-scan base-package="interceptor" />
<!-- 视图处理 -->
<bean id="viewResolver"
class="org.springframework.web.servlet.view.InternalResourceViewResolver">
    <property name="prefix" value="/WEB-INF/jsp/"></property>
    <property name="suffix" value=".jsp"></property>
</bean>
<mvc:interceptors>
    <bean class="interceptor.HelloWorldInterceptor"></bean>
    <bean class="interceptor.HelloWorldInterceptor2"></bean>
</mvc:interceptors>
```

4.3 WebRequestInterceptor

WebRequestInterceptor 中定义了三个方法，我们也是通过这三个方法来实现拦截的。这三个方法都传递了同一个参数 WebRequest，那么这个 WebRequest 是什么呢？这个 WebRequest 是 Spring 定义的一个接口，它里面的方法定义都基本跟 HttpServletRequest 一样，在 WebRequestInterceptor 中对 WebRequest 进行的所有操作都将同步到 HttpServletRequest 中，然后在当前请求中一直传递。

示例代码如下。

```
import org.springframework.ui.ModelMap;
import org.springframework.web.context.request.WebRequest;
import org.springframework.web.context.request.WebRequestInterceptor;
public class AllInterceptor implements WebRequestInterceptor {
    /**
     * 在请求处理之前执行，该方法主要是用于准备资源数据的，然后可以把它们当做请求属性放到 WebRequest 中
     */
    @Override
    public void preHandle(WebRequest request) throws Exception {
        System.out.println("AllInterceptor...............................");
```

```
            request.setAttribute("request", WebRequest.SCOPE_REQUEST);// 这个是放到 request 范围内的，所以只能在当前请求中的 request 中获取到
            request.setAttribute("session", WebRequest.SCOPE_SESSION);// 这个是放到 session 范围内的，如果环境允许的话它只能在局部的隔离的会话中访问，否则就是在普通的当前会话中可以访问
            request.setAttribute("globalSession", WebRequest.SCOPE_GLOBAL_SESSION);// 如果环境允许的话，它能在全局共享的会话中访问，否则就是在普通的当前会话中访问
    }

    /**
     * 该方法将在 Controller 执行之后，返回视图之前执行，ModelMap 表示请求 Controller 处理之后返回的 Model 对象，所以可以在
     * 这个方法中修改 ModelMap 的属性，从而达到改变返回的模型的效果。
     */
    @Override
    public void postHandle(WebRequest request, ModelMap map) throws Exception {
        // TODO Auto-generated method stub
        for (String key:map.keySet( ))
        System.out.println(key + "--");
        map.put("name3", "value3");
        map.put("name1", "name1");
    }

    /**
    * 该方法将在整个请求完成之后，也就是说在视图渲染之后进行调用，主要用于进行一些资源的释放
    */
    @Override
    public void afterCompletion(WebRequest request, Exception exception)
    throws Exception {
        // TODO Auto-generated method stub
        System.out.println(exception + "-");
    }
}
```

专家讲解

与第一种方式实现不同的是，这里的三个返回值类型都是 void。即我们不能通过此拦截器进行请求拦截。因此，我们推荐使用功能较为完整的第一种方式。

4.4 拦截器与过滤器的区别

拦截器的常见应用场景：日志记录，权限检查，性能监控，通用行为等 AOP 常用功能。

过滤器的常见应用场景：统一编码，禁止动态页面缓存，静态资源缓存，自动登陆等。

拦截器与过滤器的对比：

（1）拦截器是基于 java 的反射机制的，而过滤器是基于函数回调。

（2）拦截器不依赖与 servlet 容器，过滤器依赖与 servlet 容器。

（3）拦截器只能对 action 请求起作用，而过滤器则可以对几乎所有的请求起作用。

（4）拦截器可以访问 action 上下文、值栈里的对象，而过滤器不能访问。

（5）在 action 的生命周期中，拦截器可以多次被调用，而过滤器只能在容器初始化时被调用一次。

（6）拦截器可以获取 IoC 容器中的各个 bean，而过滤器就不行，这点很重要，在拦截器里注入一个 service，可以调用业务逻辑。

两者的本质区别：拦截器是基于 Java 的反射机制的，而过滤器是基于函数回调。从灵活性上说拦截器功能更强大些，Filter 能做的事情，他都能做，而且可以在请求前，请求后执行，比较灵活。Filter 主要是针对 URL 地址做一个编码的事情、过滤掉没用的参数、安全校验（比较泛泛的，比如登录与不登录之类），太细的话，还是建议用 interceptor。不过还是应根据不同情况选择合适的。

小结

（1）Spring MVC 也可以使用拦截器对请求进行拦截处理，用户可以自定义拦截器来实现特定的功能，自定义的拦截器必须实现 HandlerInterceptor 接口。

（2）– preHandle( )：这个方法在业务处理器处理请求之前被调用，在该方法中对用户请求 request 进行处理。如果程序员决定该拦截器对请求进行拦截处理后还要调用其他的拦截器，或者是业务处理器去进行处理，则返回 true；如果程序员决定不需要再调用其他的组件去处理请求，则返回 false。

（3）– postHandle( )：这个方法在业务处理器处理完请求后，但是 DispatcherServlet 向客户端返回响应前被调用，在该方法中对用户请求 request 进行处理。

（4）– afterCompletion( )：这个方法在 DispatcherServlet 完全处理完请求后被调用，可以在该方法中进行一些资源清理的操作。

经典面试题

（1）如何设置默认首页不拦截？

（2）拦截器与过滤器的区别是什么？

（3）Spring MVC 的拦截器，怎么设置不拦截指定的 url?

（4）如何判断 session 超时自动跳转到登录页面？

（5）如何设置静态资源不拦截？

跟我上机

（1）使用拦截器完成登录检测：在访问某些资源时（如订单页面），需要用户登录后才能查看，因此需要进行登录检测。流程如下：

①访问需要登录的资源时，由拦截器重定向到登录页面；

②如果访问的是登录页面，拦截器不应该拦截；

③用户登录成功后，往 cookie/session 添加登录成功的标识（如用户编号）；

④下次请求时，拦截器通过判断 cookie/session 中是否有该标识来决定继续流程还是到登录页面；

⑤在此拦截器还应该允许游客访问的资源。

拦截器代码如下。

```
@Override
public boolean preHandle(HttpServletRequest request, HttpServletResponse response, Object handler) throws Exception {
    //1、请求到登录页面 放行
    if(request.getServletPath( ).startsWith(loginUrl)) {
        return true;
    }
    //2、TODO 比如退出、首页等页面无需登录，即此处要放行 允许游客的请求
    //3、如果用户已经登录 放行
    if(request.getSession( ).getAttribute("username") != null) {
        // 更好的实现方式的使用 cookie
        return true;
    }
    //4、非法请求 即这些请求需要登录后才能访问
    // 重定向到登录页面
    response.sendRedirect(request.getContextPath( ) + loginUrl);
    return false;
}
```

提示：推荐能使用 servlet 规范中的过滤器 Filter 实现的功能就用 Filter 实现，因为 HandlerInteceptor 只有在 Spring Web MVC 环境下才能使用，因此 Filter 是最通用的、最先应该使用的。如登录这种拦截器最好使用 Filter 来实现。

第 5 章　Spring MVC 上传和下载

本章要点（学会后请在方框中打钩）：

- □ 掌握文件上传的配置方法和注意事项
- □ 掌握文件下载的配置方法和注意事项
- □ 掌握异步请求下的文件上传和下载
- □ 掌握使用 POI 技术文件导入导出功能

本章我们来讲讲 Spring MVC 中的文件上传和下载的几种方法。

5.1 文件上传

5.1.1 上传文件前台页面

```
<%@ page language="java" contentType="text/html; charset=utf-8" pageEncoding="utf-8"%>
<html>
<body>
<form name="serForm" action="/SpringMVC/fileUpload1" method="post"
 enctype="multipart/form-data">
<h1> 采用流的方式上传文件 </h1>
<input type="file" name="file">
        <input type="submit" value="upload" />
</form>

<form name="Form2" action="/SpringMVC/fileUpload2" method="post"
 enctype="multipart/form-data">
<h1> 采用 multipart 提供的 file.transfer 方法上传文件 </h1>
<input type="file" name="file">
        <input type="submit" value="upload" />
</form>

<form name="Form2" action="/SpringMVC/fileUpload3" method="post"
 enctype="multipart/form-data">
<h1> 使用 spring mvc 提供的类的方法上传文件 </h1>
<input type="file" name="file">
        <input type="submit"
value="upload" />
</form>
</body>
</html>
```

5.1.2 修改配置文件

```
<!-- 多部分文件上传 -->
<bean id="multipartResolver" class="org.springframework.web.multipart.commons.CommonsMultipartResolver">
    <property name="maxUploadSize" value="104857600" />
    <property name="maxInMemorySize" value="4096" />
    <property name="defaultEncoding" value="UTF-8"></property>
</bean>
```

5.1.3 编写后台控制器

采用流的方式上传文件，代码内容如下。

```
/*
 * 通过流的方式上传文件
 *
 * @RequestParam("file") 将 name=file 控件得到的文件封装成 CommonsMultipartFile 对象
 */
@RequestMapping("fileUpload1")
public String fileUpload(@RequestParam("file") CommonsMultipartFile file) throws IOException {

    // 用来检测程序运行时间
    long startTime = System.currentTimeMillis( );
    System.out.println("fileName：" + file.getOriginalFilename( ));

    try {
        // 获取输出流
        OutputStream os = new FileOutputStream("E:/" + new Date( ).getTime( ) + file.getOriginalFilename( ));
        // 获取输入流 CommonsMultipartFile 中可以直接得到文件的流
        InputStream is = file.getInputStream( );
        int temp;
        // 一个一个字节的读取并写入
        while ((temp = is.read( )) != (-1)) {
        os.write(temp);
```

```
            }
            os.flush( );
            os.close( );
            is.close( );

        } catch (FileNotFoundException e) {
            // TODO Auto-generated catch block
            e.printStackTrace( );
        }
        long endTime = System.currentTimeMillis( );
        System.out.println(" 方法一的运行时间：" + String.valueOf(endTime - startTime) + "ms");
        return "/success";
}
```

2）采用 multipart 提供的 file.transfer 方法上传文件

```
/*
 * 采用 file.Transto 来保存上传的文件
 */
@RequestMapping("fileUpload2")
public String fileUpload2(@RequestParam("file") CommonsMultipartFile file) throws IOException {
        long startTime = System.currentTimeMillis( );
        System.out.println("fileName：" + file.getOriginalFilename( ));
        String path = "E:/" + new Date( ).getTime( ) + file.getOriginalFilename( );

        File newFile = new File(path);
        // 通过 CommonsMultipartFile 的方法直接写文件（注意这个时候）
        file.transferTo(newFile);
        long endTime = System.currentTimeMillis( );
        System.out.println(" 方法二的运行时间：" + String.valueOf(endTime - startTime) + "ms");
        return "/success";
}
```

3）采用 spring 提供的上传文件的方法

```
/*
    * 采用 spring 提供的上传文件的方法
    */
   @RequestMapping("fileUpload3")
   public String springUpload(HttpServletRequest request) throws IllegalStateException, IOException {
        long startTime = System.currentTimeMillis( );
        // 将当前上下文初始化给 CommonsMutipartResolver ( 多部分解析器 )
        CommonsMultipartResolver multipartResolver = new CommonsMultipartResolver(
                request.getSession( ).getServletContext( ));
        // 检查 form 中是否有 enctype="multipart/form-data"
        if (multipartResolver.isMultipart(request)) {
            // 将 request 变成多部分 request
            MultipartHttpServletRequest multiRequest = (MultipartHttpServletRequest) request;
            // 获取 multiRequest 中所有的文件名
            Iterator iter = multiRequest.getFileNames( );
            while (iter.hasNext( )) {
                // 一次遍历所有文件
                MultipartFile file = multiRequest.getFile(iter.next( ).toString( ));
                if (file != null) {
            String path = "E:/springUpload" + file.getOriginalFilename( );
                    // 上传
                    file.transferTo(new File(path));
                }
            }
        }
        long endTime = System.currentTimeMillis( );
        System.out.println(" 方法三的运行时间：" + String.valueOf(endTime - startTime) + "ms");
        return "/success";
}
```

5.2 文件下载

```
@RequestMapping(value = "/download/{fileName}")
public void downloadFile(@PathVariable String fileName) throws IOException {
// 拼接真实路径
String realPath = getRequest( ).getServletContext( ).getRealPath("/") + "/" + fileName + ".xls";
// 读取文件
File file = new File(realPath);
if (file.exists( )) {
        OutputStream os = new BufferedOutputStream(getResponse( ).getOutputStream( ));
        try {
                getResponse( ).setContentType("application/octet-stream");
                if (getRequest( ).getHeader("User-Agent").toUpperCase( ).indexOf("MSIE") > 0) { // IE 浏览器
                        fileName = URLEncoder.encode(fileName + ".xls", "UTF-8");
                } else {
                fileName = URLDecoder.decode(fileName + ".xls");// 其他浏览器
                }
  getResponse( ).setHeader("Content-disposition",
                                "attachment; filename=" + new String(fileName.getBytes("utf-8"), "ISO8859-1")); // 指定下载的文件名
                os.write(FileUtils.readFileToByteArray(file));
                os.flush( );
        } catch (IOException e) {
                e.printStackTrace( );
        } finally {
                if (os != null) {
                        os.close( );
                }
        }
    }
}
```

小结

（1）Spring MVC 为文件上传提供了直接的支持，这种支持是通过即插即用的 MultipartResolver（接口）实现的。Spring 用 Jakarta Commons FileUpload 技术实现了一个 MultipartResolver 实现类：CommonsMultipartResovlerSpring。

（2）Spring MVC 上下文中默认没有装配 MultipartResovler，因此默认情况下不能处理文件的上传工作，如果想使用 Spring 的文件上传功能，需现在上下文中配置 MultipartResolver。

（3）Form 表单的 enctype=”multipart/form-data” 这个是上传文件必需的。

（4）applicationContext.xml 中 <bean id=”multipartResolver” class=”org.springframework.web.multipart.commons.CommonsMultipartResolver”/> 关于文件上传的配置不能少。

经典面试题

（1）什么情况下能够用到文件上传？

（2）使用 Spring MVC 进行上传文件时如何限制上传文件类型和大小？

（3）如何同时上传多个文件？

（4）负责上传文件的表单的编码类型是什么？

（5）MultipartFile 提供了获取上传文件内容、文件名等方法，通过什么方法还可以将文件存储到硬件中？

（6）文件上传表单的请求方式是什么？

（7）如何实现文件下载？

（8）如何使用 Ajax 进行异步上传？

（9）如何使用 jQuery File Upload 实现图片上传？

（10）如何使用 jQuery.form.js 插件结合 Spring MVC 实现异步上传？

跟我上机

在 Spring MVC 环境下使用 bootstrap 的 fileinput 组件完成图片文件上传功能（图 1、图 2）。

图 1　上传图片

图 2　上传图片

（2）Spring MVC 环境下用 poi 技术实现数据的导入导出 Excel 功能。

提示：

① Excel 文件数据格式如图 3 所示。

②总成绩需要计算出来。

A	B	C	D	E
姓名	语文	数学	英文	总成绩
筱静	85	48	86	219
张浩	59	69	29	
李晨	67	97	64	
杨建	79	53	79	
钱珍	89	68	58	
阮悠悠	88	78	84	

图 3　文件格式

第 6 章　Spring MVC 格式化与国际化（I18N）

本章要点(学会后请在方框中打钩)：

- □ 掌握日期格式化
- □ 掌握数字格式化
- □ 掌握货币和百分比格式化
- □ 掌握使用多种方法实现国际化

6.1 数据格式化

从 Spring3.X 开始，Spring 就提供了 Converter SPI 类型转换和 Formatter SPI 字段解析 / 格式化服务，其中 Converter SPI 实现对象与对象之间的相互转换，Formatter SPI 实现 String 与对象之间的转换，Formatter SPI 是对 Converter SPI 的封装并添加了对国际化的支持，其内部转换还是由 Converter SPI 完成。

图 1 所示为一个简单的请求与模型对象的转换流程。

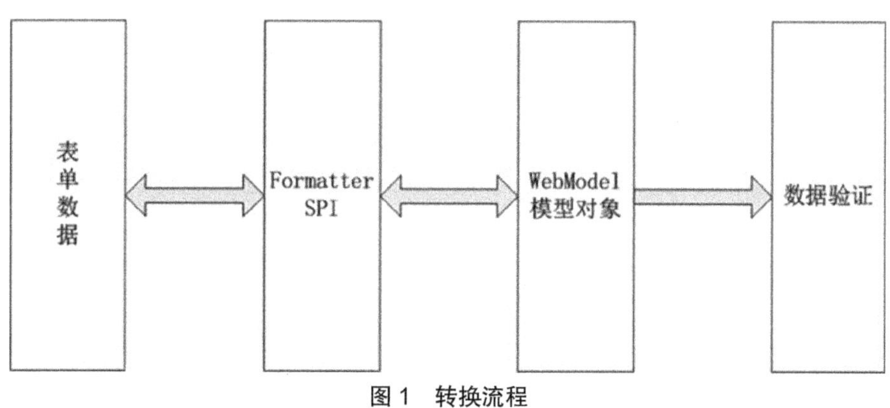

图 1 转换流程

Spring 提供了 FormattingConversionService 和 DefaultFormattingConversionService 来完成对象的解析和格式化。

Spring 内置的几种 Formatter SPI 如表 1 所示。

表 1 Spring 内置 Formatter

名称	功能
NumberFormatter	实现 Number 与 String 之间的解析与格式化
CurrencyFormatter	实现 Number 与 String 之间的解析与格式化（带货币符号）
PercentFormatter	实现 Number 与 String 之间的解析与格式化（带百分数符号）
DateFormatter	实现 Date 与 String 之间的解析与格式化
NumberFormatAnnotation-FormatterFactory	@NumberFormat 注解，实现 Number 与 String 之间的解析与格式化，可以通过指定 Style 来指示要转换的格式（Style.Number/Style.Currency/Style.Percent），当然也可以指定 pattern（如 pattern="#.##"（保留 2 位小数）），这样 pattern 指定的格式会覆盖掉 Style 指定的格式
JodaDateTimeFormatAnnota-tionFormatterFactory	@DateTimeFormat 注解，实现日期类型与 String 之间的解析与格式化这里的日期类型包括 Date、Calendar、Long 以及 Joda 的日期类型。必须在项目中添加 Joda-Time 包

6.1.1 使用 CurrencyFormatter 和 DateFormatter 格式化货币和日期

首先把 Joda-Time 包添加到之前的项目中，这里用的是 joda-time-2.9.jar，在 views 文件夹下添加一个 formattest.jsp 视图，代码内容如下。

```
<%@ page language="java" contentType="text/html; charset=UTF-8"    pageEncoding="UTF-8"%>
<html>
<body>
    money:<br/>${contentModel.money}<br/>
    date:<br/>${contentModel.date}<br/>
</body>
</html>
```

首先我们直接用 Formatter 来做演示，在 com.iss.models 包中添加 FormatModel.java 文件，内容如下。

```
package com.iss.models;
public class FormatModel{
    private String money;
private String date;
//setter 和 getter 方法略
}
```

在 com.iss.controllers 包中添加 FormatController.java 文件，内容如下。

```
package com.iss.controllers;
//import 导包略
@Controller
@RequestMapping(value = "/format")
public class FormatController {

    @RequestMapping(value="/test", method = {RequestMethod.GET})
    public String test(Model model) throws NoSuchFieldException, SecurityException{
        if(!model.containsAttribute("contentModel")){
            FormatModel formatModel=new FormatModel( );
            CurrencyFormatter currencyFormatter = new CurrencyFormatter( );
            currencyFormatter.setFractionDigits(2);// 保留 2 位小数
            currencyFormatter.setRoundingMode(RoundingMode.HALF_UP);// 向（距离）最近的一边舍入，如果两边（的距离）是相等的则向上舍入（四舍五入）
```

```
                DateFormatter dateFormatter=new DateFormatter( );
                dateFormatter.setPattern("yyyy-MM-dd HH:mm:ss");
                Locale locale=LocaleContextHolder.getLocale( );
                formatModel.setMoney(currencyFormatter.print(12345.678, locale));
                formatModel.setDate(dateFormatter.print(new Date( ), locale));
                model.addAttribute("contentModel", formatModel);
            }
            return "formattest";
        }
}
```

浏览器运行测试（见图 2）。

money:
¥12,345.68
date:
2017-06-21 22:56:23

图 2　浏览器运行测试结果

更改浏览器首选语言，刷新页面（见图 3）。

money:
$12,345.68
date:
2017-06-21 22:59:19

改成英语

图 3　刷新页面

6.1.2　使用 DefaultFormattingConversionService 进行格式化

这次用 DefaultFormattingConversionService 来做演示，把 FormatController.java 文件的内容改为如下内容。

```
package com.iss.controllers;

@Controller
@RequestMapping(value = "/format")
public class FormatController {
    @RequestMapping(value="/test", method = {RequestMethod.GET})
    public String test(Model model) throws NoSuchFieldException, SecurityException{
        if(!model.containsAttribute("contentModel")){
            FormatModel formatModel=new FormatModel( );
```

```
                CurrencyFormatter currencyFormatter = new CurrencyFormatter( );
                currencyFormatter.setFractionDigits(2);// 保留 2 位小数
                currencyFormatter.setRoundingMode(RoundingMode.HALF_UP);// 向（距离）最近的一边舍入，如果两边（的距离）是相等的则向上舍入（四舍五入）
                DateFormatter dateFormatter=new DateFormatter( );
                dateFormatter.setPattern("yyyy-MM-dd HH:mm:ss");
                DefaultFormattingConversionService conversionService = new DefaultFormattingConversionService( );
                conversionService.addFormatter(currencyFormatter);
                conversionService.addFormatter(dateFormatter);
                formatModel.setMoney(conversionService.convert(12345.678, String.class));
                formatModel.setDate(conversionService.convert(new Date( ), String.class));
                model.addAttribute("contentModel", formatModel);
            }
            return "formattest";
        }
}
```

这次没有了 Locale locale=LocaleContextHolder.getLocale( ); 再次运行测试并更改语言后刷新，可以看到与第一种方法截图同样的效果，说明 DefaultFormattingConversionService 会自动根据浏览器请求的信息返回相应的格式。

6.1.3 使用注解进行格式化

上面只是对内置的格式化转换器做了一下演示，实际项目中肯定不会这么用的，下面就介绍一下基于注解的格式化。首先把 FormatModel.java 改为如下内容。

```
package com.iss.models;
import java.util.Date;
import org.springframework.format.annotation.DateTimeFormat;
import org.springframework.format.annotation.NumberFormat;
import org.springframework.format.annotation.NumberFormat.Style;

public class FormatModel{
    @NumberFormat(style=Style.CURRENCY)private double money;
    @DateTimeFormat(pattern="yyyy-MM-dd HH:mm:ss")
    private Date date;
    //Setter 和 getter 方法略
}
```

注意：这里的 money 和 date 不再是 String 类型，而是它们自己本来的类型。

把 FormatController.java 改为如下内容。

```
package com.iss.controllers;

@Controller
@RequestMapping(value = "/format")
public class FormatController {

        @RequestMapping(value="/test", method = {RequestMethod.GET})
        public String test(Model model) throws NoSuchFieldException, SecurityException{
                if(!model.containsAttribute("contentModel")){
                        FormatModel formatModel=new FormatModel( );
                        formatModel.setMoney(12345.678);
                        formatModel.setDate(new Date( ));
                        model.addAttribute("contentModel", formatModel);
                }
         return "formattest";
                }
}
```

注意：这里代码里面只有赋值已经没有格式化的内容了。

更改视图 formattest.jsp 的内容如下。

```
<%@ page language="java" contentType="text/html; charset=UTF-8"
    pageEncoding="UTF-8"%>
<%@taglib prefix="spring" uri="http://www.springframework.org/tags" %>
<html>
<head>
<meta http-equiv="Content-Type" content="text/html; charset=UTF-8">
<title>Insert title here</title>
</head>
<body>
    money:<br/>
    <spring:eval expression="contentModel.money"></spring:eval><br/>
    date:<br/>
    <spring:eval expression="contentModel.date"></spring:eval><br/>
</body>
</html>
```

注意：这里需要添加引用 <%@taglib prefix="spring"。

uri="http://www.springframework.org/tags"%>，并用 spring:eval 来绑定要显示的值。

运行测试更改浏览器语言然后刷新页面依然可以看到以第一种方法截图相同的效果，证明注解有效。

6.2 国际化 (I18N)

上一节我们讲了数据的格式化显示，Spring 在做格式化展示的时候已经做了国际化处理，那么如何将我们网站的其他内容（如菜单、标题等）做国际化处理呢？本节要讲的内容就是国际化（I18N）。所谓国际化就是支持多种语言，web 应用在不同的浏览环境中可以显示出不同的语言，比如汉语、英语等。

I18N（其来源是英文单词 internationalization 的首末字符 i 和 n，18 为中间的字符数）是“国际化”的简称。在资讯领域，国际化（i18n）指让产品（出版物、软件、硬件等）无须做大的改变就能够适应不同的语言和地区的需要。对程序来说，在不修改内部代码的情况下，能根据不同语言及地区显示相应的界面。 在全球化的时代，国际化尤为重要，因为产品的潜在用户可能来自世界的各个角落。通常与 i18n 相关的还有 L10n（“本地化”的简称）。

6.2.1 基于浏览器请求的国际化实现

首先配置我们项目的 springservlet-config.xml 文件添加的内容如下。

```
<bean id="messageSource" class="org.springframework.context.support.ResourceBundleMessageSource">
    <!-- 国际化信息所在的文件名 -->
    <property name="basename" value="messages" />
    <!-- 如果在国际化资源文件中找不到对应代码的信息，就用这个代码作为名称 -->
    <property name="useCodeAsDefaultMessage" value="true" />
</bean>
```

在 com.iss.controllers 包中添加 GlobalController.java 内容如下。

```
package com.iss.controllers;

@Controller
@RequestMapping(value = "/global")
public class GlobalController {
  @RequestMapping(value="/test", method = {RequestMethod.GET})
    public String test(HttpServletRequest request,Model model){
        if(!model.containsAttribute("contentModel")){
```

```
            // 从后台代码获取国际化信息
            RequestContext requestContext = new RequestContext(request);
            model.addAttribute("money", requestContext.getMessage("money"));
            model.addAttribute("date", requestContext.getMessage("date"));
            FormatModel formatModel=new FormatModel( );
            formatModel.setMoney(12345.678);
            formatModel.setDate(new Date( ));
            model.addAttribute("contentModel", formatModel);
        }
        return "globaltest";
    }
}
```

在项目中的源文件夹 resources 中添加 messages.properties、messages_zh_CN.properties、messages_en_US.properties 三个文件，其中 messages.properties、messages_zh_CN.properties 里面的 "money", "date", 为中文，messages_en_US.properties 里面的为英文。

在 views 文件夹中添加 globaltest.jsp 视图，内容如下。

```
<%@ page language="java" contentType="text/html; charset=UTF-8"
    pageEncoding="UTF-8"%>
<%@taglib prefix="spring" uri="http://www.springframework.org/tags" %>
<html>
<body>
  下面展示的是后台获取的国际化信息：<br/>
    ${money}<br/>
    ${date}<br/>

    下面展示的是视图中直接绑定的国际化信息：<br/>
    <spring:message code="money"/>:<br/>
    <spring:eval expression="contentModel.money"></spring:eval><br/>
    <spring:message code="date"/>:<br/>
    <spring:eval expression="contentModel.date"></spring:eval><br/>
    </body>
</html>
```

运行测试，更改浏览器语言顺序，刷新页面，查看结果。

6.2.2 基于 Session 的国际化实现

在项目的 springservlet-config.xml 文件添加的内容如下（第一种时添加的内容要保留）。

```
<mvc:interceptors>
    <!-- 国际化操作拦截器 如果采用基于（请求 /Session/Cookie) 则必需配置 -->
    <bean class="org.springframework.web.servlet.i18n.LocaleChangeInterceptor" />
</mvc:interceptors>

<bean id="localeResolver" class="org.springframework.web.servlet.i18n.SessionLocaleResolver" />
```

更改 globaltest.jsp 视图为如下内容。

```
<%@ page language="java" contentType="text/html; charset=UTF-8" pageEncoding="UTF-8"%>
<%@taglib prefix="spring" uri="http://www.springframework.org/tags" %>
<body>
     <a href="test?langType=zh"> 中文 </a> | <a href="test?langType=en"> 英文 </a><br/>

    下面展示的是后台获取的国际化信息：<br/>
    ${money}<br/>
    ${date}<br/>

    下面展示的是视图中直接绑定的国际化信息：<br/>
    <spring:message code="money"/>:<br/>
    <spring:eval expression="contentModel.money"></spring:eval><br/>
    <spring:message code="date"/>:<br/>
    <spring:eval expression="contentModel.date"></spring:eval><br/>
</body>
</html>
```

更改 GlobalController.java 文件为如下内容。

```
package com.iss.controllers;

@Controller
@RequestMapping(value = "/global")
public class GlobalController {

    @RequestMapping(value="/test", method = {RequestMethod.GET})
```

```
    public String test(HttpServletRequest request,Model model, @RequestParam(value="langType", defaultValue="zh") String langType){
        if(!model.containsAttribute("contentModel")){
            if(langType.equals("zh")){
                Locale locale = new Locale("zh", "CN");
request.getSession( ).setAttribute(SessionLocaleResolver.LOCALE_SESSION_ATTRIBUTE_NAME,locale);
            }
            else if(langType.equals("en")){
                Locale locale = new Locale("en", "US");
request.getSession( ).setAttribute(SessionLocaleResolver.LOCALE_SESSION_ATTRIBUTE_NAME,locale);
            }
            else
request.getSession( ).setAttribute(SessionLocaleResolver.LOCALE_SESSION_ATTRIBUTE_NAME,LocaleContextHolder.getLocale( ));
            // 从后台代码获取国际化信息
            RequestContext requestContext = new RequestContext(request);
            model.addAttribute("money", requestContext.getMessage("money"));
            model.addAttribute("date", requestContext.getMessage("date"));
            FormatModel formatModel=new FormatModel( );
            formatModel.setMoney(12345.678);
            formatModel.setDate(new Date( ));
            model.addAttribute("contentModel", formatModel);
        }
        return "globaltest";
    }
}
```

6.2.3 基于 Cookie 的国际化实现

把实现第二种方法时在项目的 springservlet-config.xml 文件中添加的注释掉，并添加以下内容。

```
<bean id="localeResolver"
class="org.springframework.web.servlet.i18n.SessionLocaleResolver" />
```

```
<bean id="localeResolver"
class="org.springframework.web.servlet.i18n.CookieLocaleResolver" />
```

更改 GlobalController.java 为如下内容。

```
package com.iss.controllers;
@Controller
@RequestMapping(value = "/global")
public class GlobalController {

    @RequestMapping(value="/test", method = {RequestMethod.GET})
        public String test(HttpServletRequest request, HttpServletResponse response, Model model, @RequestParam(value="langType", defaultValue="zh") String langType){
        if(!model.containsAttribute("contentModel")){
            if(langType.equals("zh")){
                Locale locale = new Locale("zh", "CN");
                (new CookieLocaleResolver( )).setLocale (request, response, locale);
            }
            else if(langType.equals("en")){
                Locale locale = new Locale("en", "US");
                    (new CookieLocaleResolver( )).setLocale (request, response, locale);
            }
            else
                (new CookieLocaleResolver( )).setLocale (request, response, LocaleContextHolder.getLocale( ));
            // 从后台代码获取国际化信息
            RequestContext requestContext = new RequestContext(request);
            model.addAttribute("money", requestContext.getMessage("money"));
            model.addAttribute("date", requestContext.getMessage("date"));
            FormatModel formatModel=new FormatModel( );
            formatModel.setMoney(12345.678);
            formatModel.setDate(new Date( ));
            model.addAttribute("contentModel", formatModel);
        }
        return "globaltest";
    }
}
```

关于 <bean id="localeResolver" class="org.springframework.web.servlet.i18n.CookieLocaleResolver" />3 个属性的说明（可以都不设置而用其默认值）。

```
<bean id="localeResolver"
  class="org.springframework.web.servlet.i18n.CookieLocaleResolver">
    <!-- 设置 cookieName 名称，可以根据名称通过 js 来修改设置，也可以像上面演示的那样修改设置，默认的名称为 类名 +LOCALE（即：org.springframework.web.servlet.i18n.CookieLocaleResolver.LOCALE-->
    <property name="cookieName" value="lang"/>
    <!-- 设置最大有效时间，如果是 -1，则不存储，浏览器关闭后即失效，默认为 Integer.MAX_INT-->
    <property name="cookieMaxAge" value="100000">
    <!-- 设置 cookie 可见的地址，默认是"/"即对网站所有地址都是可见的，如果设为其它地址，则只有该地址或其后的地址才可见 -->
    <property name="cookiePath" value="/">
</bean>
```

6.2.4 基于 URL 请求的国际化的实现

首先添加一个类，内容如下。

```
public class MyAcceptHeaderLocaleResolver extends AcceptHeaderLocaleResolver {
        private Locale myLocal;
        public Locale resolveLocale(HttpServletRequest request) {
                return myLocal;
        }
        public void setLocale(HttpServletRequest request, HttpServletResponse response, Locale locale) {
                myLocal = locale;
        }
}
```

然后把实现第二种方法时在项目的 springservlet-config.xml 文件中添加的内容注释掉，并添加以下内容。

```
<bean id="localeResolver"
class="org.springframework.web.servlet.i18n.SessionLocaleResolver" />
```

```
<bean id="locale Resolver" class="com.iss.resoler.MyAcceptHeaderLocaleResolver"/>
```

6.3 综合实例演示

该项目中采用 Spring MVC 的框架，采用基于 session 的动态切换，实现动态切换中文、英文、韩文，其实就是把中文翻译成其他语言显示。

Spring MVC 国际化包括两个方面，一个是前台页面的国际化，spring 有自己的标签可以去实现，非常方便，另一个是后台 java 代码中涉及中文的国际化。

6.3.1 项目总体结构

项目总体结构如图 4 所示。

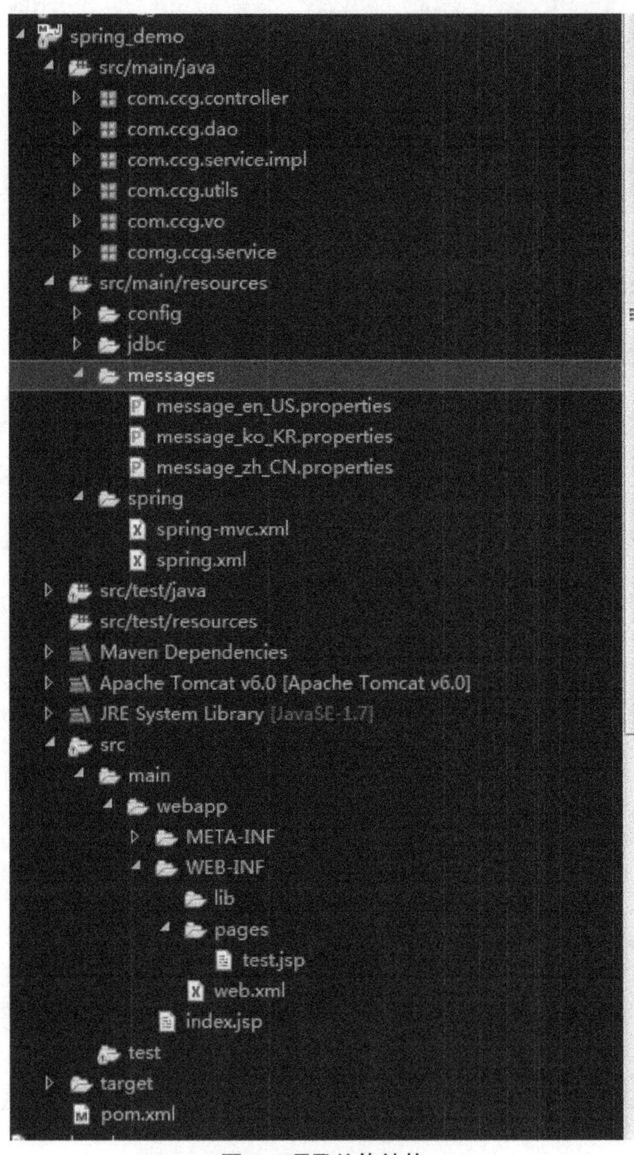

图 4　项目总体结构

有关 bean 的主要配置在 spring-mvc.xml 里配置，messages 文件夹里放的是需要翻译的，内容格式如下：key=value 的格式（见图 5 至图 7）。

英文：

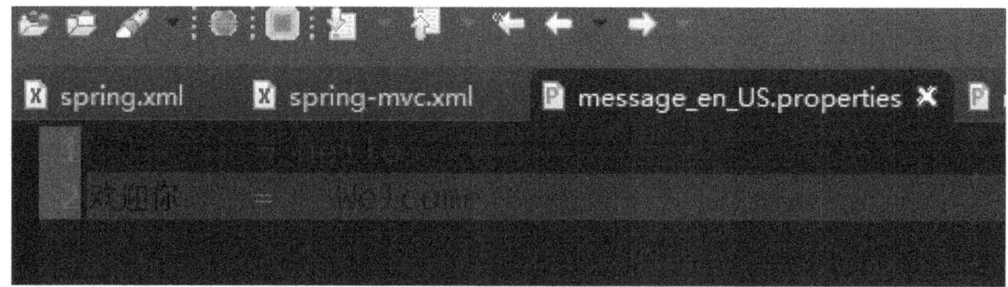

图 5　英文界面

中文：

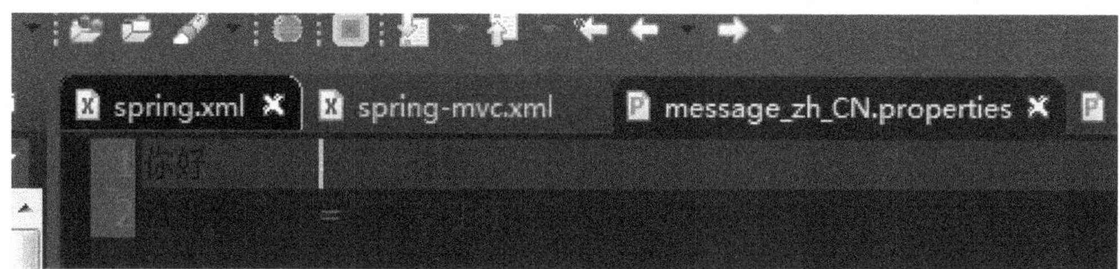

图 6　中文界面

韩文：

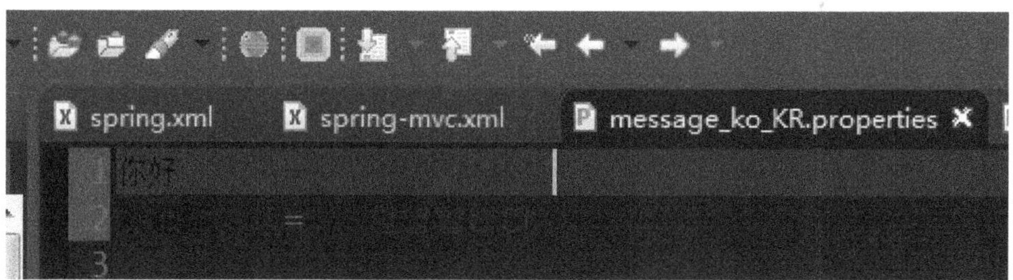

图 7　韩文界面

专家讲解

需要注意的是配置文件里面涉及的中文需要转成 unicode 编码，否则翻译后会出现乱码。

6.3.2 在 spring-mvc 加入以下配置

```
<!-- 国际化资源配置，资源文件绑定器 -->
        <bean id="messageSource" class="org.springframework.context.support.ReloadableResourceBundleMessageSource">
        <!-- 国际化资源文件配置，指定 properties 文件存放位置 -->
        <property name="basename" value="classpath:messages/message" />
         <!-- 如果在国际化资源文件中找不到对应代码的信息，就用这个代码作为名称 -->
         <property name="useCodeAsDefaultMessage" value="true" />
    </bean>
    <!-- 动态切换国际化，国际化放在 session 中 -->
     <bean id="localeResolver" class="org.springframework.web.servlet.i18n.SessionLocaleResolver"></bean>
    <mvc:interceptors>
        <!-- 国际化操作拦截器 如果采用基于（请求 /Session/Cookie) 则必需配置 -->
        <bean class="org.springframework.web.servlet.i18n.LocaleChangeInterceptor">
            <!-- 通过这个参数来决定获取那个配置文件 -->
          <property name="paramName" value="language" />
        </bean>
   </mvc:interceptors>
```

需要注意的是 basename 的值需要配置到 messages/message 这一级别（见图 7）。

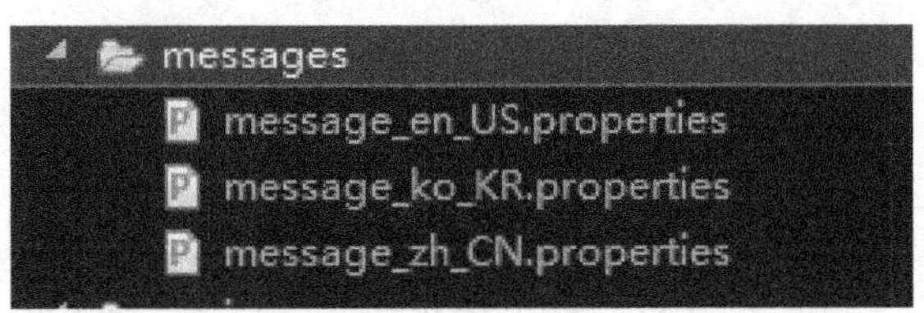

图 7　配置文件

6.3.3 前台页面实现

```
<%@ page language="java" contentType="text/html; charset=UTF-8"
        pageEncoding="UTF-8"%>
<%@taglib prefix="spring" uri="http://www.springframework.org/tags" %>
<%@taglib prefix="mvc" uri="http://www.springframework.org/tags/form" %>
<html>
<head>
```

```
<meta http-equiv="Content-Type" content="text/html; charset=UTF-8">
<title>Insert title here</title>
</head>
<body>
选择语言：<a href="?language=zh_CN"> 中文 </a> | <a href="?language=en_US"> 英文 </a> | <a href="?language=ko_KR"> 韩文 </a>
<br></br>
这里展示选择对应语言后的“你好”的翻译（前台标签翻译）：
<spring:message code=" 你好 " />
<br></br>
这里展示选择对应语言后的“欢迎你”的翻译（后台代码翻译）：${welcome}
</body>
</html>
```

专家讲解

需要注意的是页面引入 spring 的标签，code 这里写的是中文，对应那三个翻译文件里的 key 如果 code 在翻译文件里没有找到，则使用 code 里面的内容，这个配置在 Spring MVC 里指定了。

6.3.4 后台代码的实现

```
package com.iss.controller;
@Controller
@RequestMapping("test")
public class TestController {
    @RequestMapping("view")
    public ModelAndView view(HttpServletRequest request, HttpServletResponse response){
        //spring 翻译使用 req.getmessage( ) 方法
        RequestContext req = new RequestContext(request);
        ModelAndView model = new ModelAndView("test");
        model.addObject("welcome", req.getMessage(" 欢迎你 "));
        return model;
    }
}
```

6.3.5 效果展示

效果展示如图 8 至图 10。

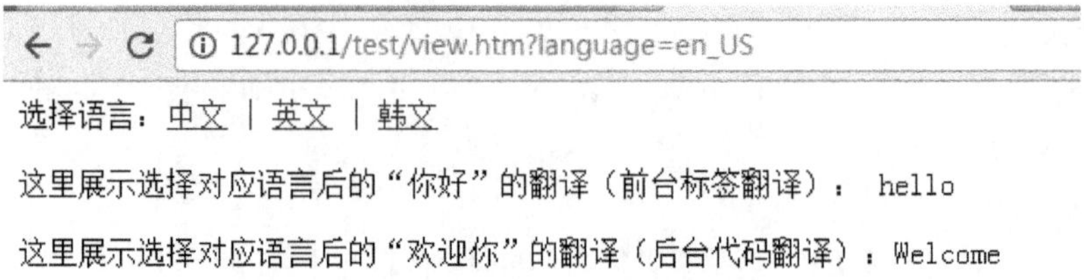

127.0.0.1/test/view.htm?language=en_US

选择语言：中文 | 英文 | 韩文

这里展示选择对应语言后的“你好”的翻译（前台标签翻译）： hello

这里展示选择对应语言后的“欢迎你”的翻译（后台代码翻译）：Welcome

图 8 英文效果展示

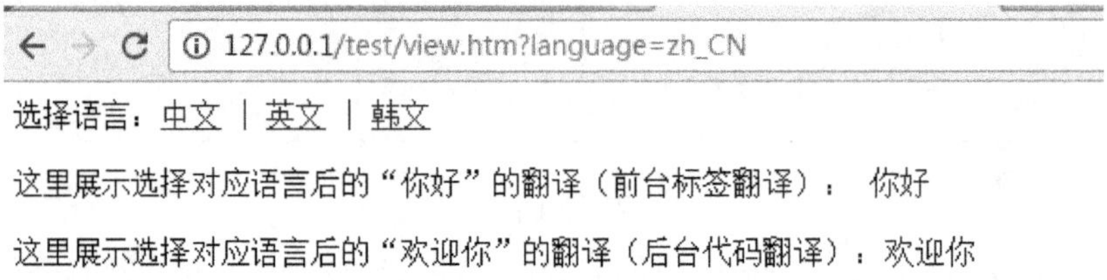

127.0.0.1/test/view.htm?language=zh_CN

选择语言：中文 | 英文 | 韩文

这里展示选择对应语言后的“你好”的翻译（前台标签翻译）： 你好

这里展示选择对应语言后的“欢迎你”的翻译（后台代码翻译）：欢迎你

图 9 中文效果展示

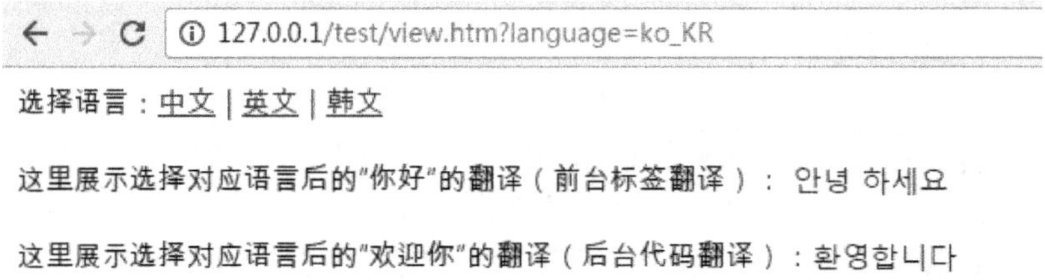

127.0.0.1/test/view.htm?language=ko_KR

选择语言：中文 | 英文 | 韩文

这里展示选择对应语言后的“你好”的翻译（前台标签翻译）： 안녕 하세요

这里展示选择对应语言后的“欢迎你”的翻译（后台代码翻译）：환영합니다

图 10 韩文效果展示

小结

（1）总结一下 Spring MVC 的国际化的关键点。

（2）指定 Spring 国际化需要翻译的文件位置，需要注意的是路径一定要写完整。

（3）指定 Spring 国际化时的参数，上文中用到是 language，配置好之后，spring 会根据请求中该参数的值找到对应的配置文件。

（4）前台页面使用 Spring 标签， code 对应配置文件中的 key，推荐使用中文，万一对应不上的时候默认显示 code 里面的内容（需要在 spring-mvc.xml 里指定配置）。

（5）后台代码国际化使用 RequestContext 对象的 getMessage 方法，返回 String。

经典面试题

（1）Spring MVC 中如何格式化数据？
（2）如何实现 Spring MVC 国际化设置？
（3）Spring MVC 的国际化，动态设置默认语言如何实现？

跟我上机

使用 Spring MVC 进行国际化文件配置，根据下拉框选择中英文页面，实现效果如图 11 所示。

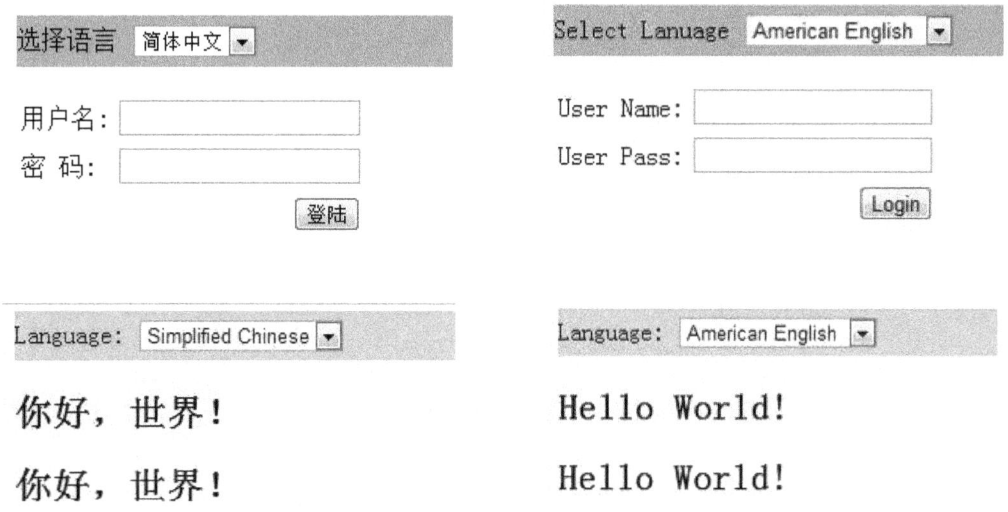

图 11 效果图

第 7 章　Spring MVC 异常处理

本章要点(学会后请在方框中打钩):

- □ 掌握异常处理方式
- □ 掌握异常处理的思路和设计方法
- □ 掌握 Spring MVC 中自带的简单异常处理器
- □ 掌握用户自定义异常处理器
- □ 掌握使用注解实现异常处理

在 Java EE 项目的开发中，不管是对底层的数据库操作过程，还是业务层的处理过程，还是控制层的处理过程，都不可避免会遇到各种可预知的、不可预知的异常需要处理。每个过程都单独处理异常，系统的代码耦合度高，工作量大且不好统一，维护的工作量也很大。

那么，能不能将所有类型的异常处理从各处理过程解耦出来，既保证相关处理过程的功能较单一，也实现了异常信息的统一处理和维护。答案是肯定的。下面将介绍使用 Spring MVC 统一处理异常的解决和实现过程。

7.1 Spring MVC 的处理异常方式

Spring MVC 的处理异常有以下 3 种方式。

（1）使用 Spring MVC 提供的简单异常处理器 SimpleMappingExceptionResolver。

（2）实现 Spring 的异常处理接口 HandlerExceptionResolver 自定义自己的异常处理器。

（3）使用 @ExceptionHandler 注解实现异常处理。

7.2 异常处理机制

首先来看一下在 Spring MVC 中，异常处理的思路，如图 1 所示。

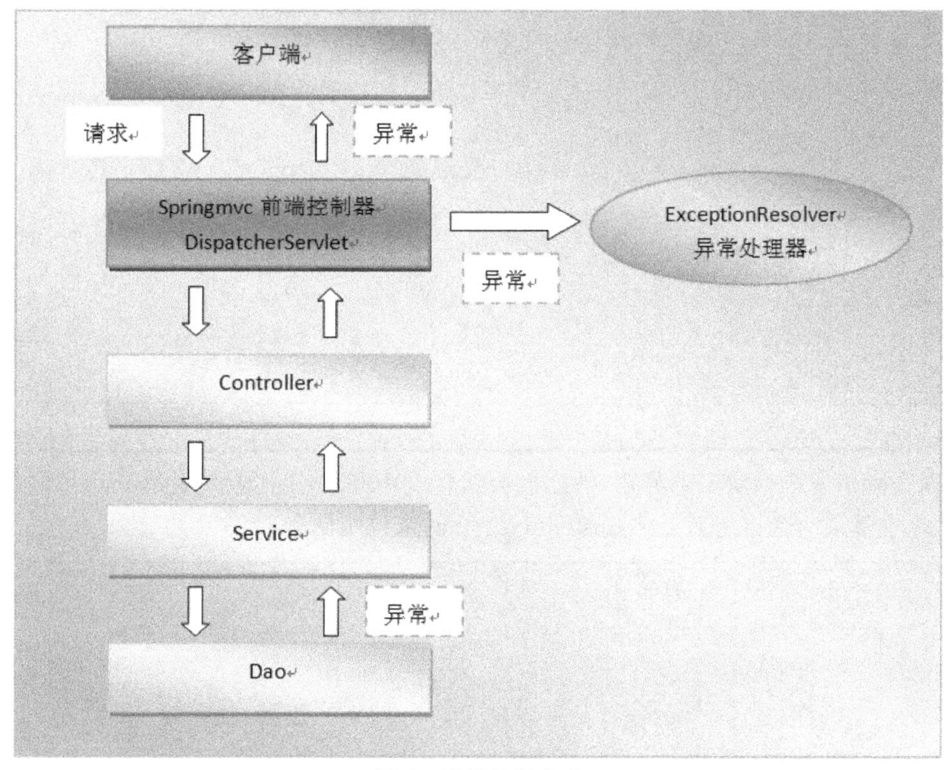

图 1　异常处理方式

如图 1 所示，系统的 dao、service、controller 出现异常都通过 throws Exception 向上抛出，最后由 Spring MVC 前端控制器交由异常处理器进行异常处理。Spring MVC 提供全局异常处理器（一个系统只有一个异常处理器）进行统一异常处理。明白了 Spring MVC 中的异常处理机制以后，下面就开始分析 Spring MVC 中的异常处理。

7.3 使用自带的简单异常处理器

Spring MVC 中自带了一个异常处理器叫 SimpleMappingExceptionResolver，该处理器实现了 HandlerExceptionResolver 接口，全局异常处理器都需要实现该接口。我们要使用这个自带的异常处理器，首先得在 springmvc.xml 文件中配置该处理器。

```
<!-- springmvc 提供的简单异常处理器 -->
<bean class="org.springframework.web.servlet.handler.SimpleMappingExceptionResolver">
    <!-- 定义默认的异常处理页面 -->
    <property name="defaultErrorView" value="/WEB-INF/jsp/error.jsp"/>
    <!-- 定义异常处理页面用来获取异常信息的变量名，也可不定义，默认名为 exception -->
    <property name="exceptionAttribute" value="ex"/>
    <!-- 定义需要特殊处理的异常，这是重要点 -->
    <property name="exceptionMappings">
        <props>
            <prop key="ssm.exception.CustomException">/WEB-INF/jsp/custom_error.jsp</prop>
        </props>
        <!-- 还可以定义其他的自定义异常 -->
    </property>
</bean>
```

从上面的配置来看，最重要的是要配置特殊处理的异常，这些异常一般都是我们自定义的，根据实际情况来自定义的异常，然后也会跳转到不同的错误显示页面，显示不同的错误信息。这里就用一个自定义异常 CustomException 来说明问题，定义如下。

```
// 定义一个简单的异常类
public class CustomException extends Exception {
    // 异常信息
    public String message;
    public CustomException(String message) {
```

```
        super(message);
        this.message = message;
    }
    public String getMessage( ) {
        return message;
    }
    public void setMessage(String message) {
        this.message = message;
    }
}
```

接下来就是写测试程序了，还是使用查询的例子，如图 2 所示。

```
@RequestMapping("/editItems")
public String editItems(Model model,
        @RequestParam(value = "id", required = true) Integer items_id)
        throws Exception {
    // 根据id查询对应的Items
    ItemsCustom itemsCustom = itemsService.findItemsById(items_id);

    if(itemsCustom == null) {
        throw new CustomException("修改的商品信息不存在！");
    }

    model.addAttribute("itemsCustom", itemsCustom);

    return "/WEB-INF/jsp/items/editItems.jsp";
}
```

如果查询的商品不存在则抛出我们自定义的异常

如果存在则正常往下执行

图 2　测试程序

然后在前台输入 url 来测试：http://localhost:8080/SpringMVC_Study/editItems.action?id=11，故意传一个 id 为 11，数据库中没有 id 为 11 的项，所以肯定查不到。这样它就会抛出自定义的异常，然后被上面配置的全局异常处理器捕获并执行，跳转到指定的页面，然后显示一下该商品不存在即可。所以这个流程是很清晰的。

从上面的过程可知，使用 SimpleMappingExceptionResolver 进行异常处理，具有集成简单、有良好的扩展性（可以任意增加自定义的异常和异常显示页面）、对已有代码没有入侵性等优点，但该方法仅能获取到异常信息，若在出现异常时，对需要获取除异常以外的数据的情况不适用。

7.4　自定义全局异常处理器

全局异常处理器的处理思路如下。

（1）解析出异常类型。

（2）如果该异常类型是系统自定义的异常，直接取出异常信息，在错误页面展示。

（3）如果该异常类型不是系统自定义的异常，构造一个自定义的异常类型（信息为“未知错误”）。

Spring MVC 提供一个 HandlerExceptionResolver 接口，自定义全局异常处理器必须要实现这个接口，如下。

```
public class CustomExceptionResolver implements HandlerExceptionResolver {
        @Override
        public ModelAndView resolveException(HttpServletRequest request,
                    HttpServletResponse response, Object handler, Exception ex) {
            ex.printStackTrace( );
            CustomException customException = null;
            // 如果抛出的是系统自定义的异常则直接转换
            if(ex instanceof CustomException) {
                customException = (CustomException) ex;
            } else {
                // 如果抛出的不是系统自定义的异常则重新构造一个未知错误异常
                // 这里我就也有 CustomException 省事了，实际中应该要再定义一个新的异常
                customException = new CustomException(" 系统未知错误 ");
            }
            // 向前台返回错误信息
            ModelAndView modelAndView = new ModelAndView( );
            modelAndView.addObject("message", customException.getMessage( ));
            modelAndView.setViewName("/WEB-INF/jsp/error.jsp");
            return modelAndView;
        }
}
```

全局异常处理器中的逻辑很清楚，这里就不再多说了，然后就是在 springmvc.xml 中配置这个自定义的异常处理器。

```
<!-- 自定义的全局异常处理器
只要实现 HandlerExceptionResolver 接口就是全局异常处理器 -->
<bean class="ssm.exception.CustomExceptionResolver"></bean>
```

然后就可以使用上面那个测试用例再次测试了。可以看出在自定义的异常处理器中能获取导致出现异常的对象，有利于提供更详细的异常处理信息。一般用这种自定义的全局异常处理器比较多。

7.5 使用 @ExceptionHandler 实现异常处理

首先写个 BaseController 类，并在类中使用 @ExceptionHandler 注解声明异常处理的方法，如下。

```
public class BaseController {
    @ExceptionHandler
    public String exp(HttpServletRequest request, Exception ex) {
    // 异常处理
    //......
    }
}
```

然后将所有需要异常处理的 Controller 都继承这个 BaseController，虽然从执行来看，不需要配置什么东西，但是代码有侵入性，需要异常处理的 Controller 都要继承它才行。

小结

Spring MVC 集成异常处理的 3 种方式都可以达到统一异常处理的目标。比较 3 种方式的优缺点，若只需要简单的集成异常处理，推荐使用 SimpleMappingExceptionResolver；若需要集成的异常处理能够更具个性化，提供给用户更详细的异常信息，推荐自定义实现 HandlerExceptionResolver 接口的方式；若不喜欢 Spring 配置文件或要实现“零配置”，且能接受对原有代码的适当入侵，则建议使用 @ExceptionHandler 注解方式。

经典面试题

（1）通常情况下 Spring MVC 异常处理有几种方法？

（2）什么是 Spring MVC 异常处理机制？

（3）Spring MVC 如何配置异常处理？

（4）Spring MVC 注解配置异常时用哪个注解配置捕获异常？

（5）如何使用 @ControllerAdvice 注解来实现全局异常处理？

跟我上机

使用 Spring MVC 异常处理机制和注解和配置文件方式，实现请求资源请求不到显示 404 页面的功能（见图 3）。

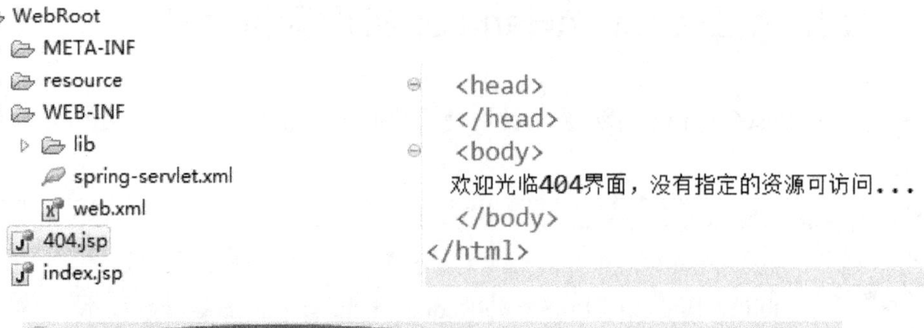

localhost:8080/SpringMVCTest/test/ 控制层找不到该地址对应的资源

最常访问 now 营业厅 测试系统 SpringMVCTest

欢迎光临404界面，没有指定的资源可访问...

图 3　示意图

第 3 篇　MyBatis 持久层框架

学习目标：

- □ 了解 MyBatis 介绍
- □ 掌握 MyBatis 基本配置
- □ 掌握 configuration.xml 文件配置详解
- □ 掌握日志配置
- □ 精通 MyBatis 基本增改删查（CRUD）
- □ 了解 MyBatis 映射文件
- □ 掌握动态查询
- □ 掌握基于注解配置实现
- □ 掌握 MyBatis 分页查询
- □ 掌握 MyBatis 调用存储过程
- □ 掌握 MyBatis 缓存机制

第 1 章　MyBatis 介绍

本章要点（学会后请在方框中打钩）：

□ 了解 MyBatis 的前世今生

□ 了解 MyBatis 的优点

□ 了解 MyBatis 与传统 JDBC 相比的优势

□ 了解 JDBC 与 MyBatis 的直观对比

□ 了解 MyBatis 和 Hibernate 的对比

□ 了解 MyBatis 的工作流程

1.1 MyBatis 的前世今生

MyBatis 的前身是 iBatis，iBatis 本是由 Clinton Begin 开发的，后来捐给 Apache 基金会，成立了 iBatis 开源项目。2010 年 5 月该项目由 Apahce 基金会迁移到了 Google Code，并改名为 MyBatis。

MyBatis 是一个数据持久层 (ORM) 框架。在实体类和 SQL 语句之间建立了映射关系，是一种半自动化的 ORM 实现。

MyBatis 是一个支持普通 SQL 查询、存储过程和高级映射的优秀持久层框架。MyBatis 消除了几乎所有的 JDBC 代码和参数的手工设置以及对结果集的检索封装。MyBatis 可以使用简单的 XML 或注解用于配置和原始映射，将接口和 Java 的 POJO（Plain Old Java Objects，普通的 Java 对象）映射成数据库中的记录。

1.2 MyBatis 的优点

（1）基于 SQL 语法，简单易学。

（2）能了解底层组装过程。

（3）SQL 语句封装在配置文件中，便于统一管理与维护，降低了程序的耦合度。

（4）程序调试方便。

1.3 与传统 JDBC 相比的优势

（1）减少了 61% 的代码量。

（2）最简单的持久化框架。

（3）架构级性能增强。

（4）SQL 代码从程序代码中彻底分离，可重用。

（5）增强了项目中的分工。

（6）增强了移植性。

1.4 JDBC 与 MyBatis 的直观对比

```
1      Class.forName("oracle.jdbc.driver.OracleDriver");
2      Connection conn= DriverManager.getConnection(url,user,password);
3      java.sql.PreparedStatement  st = conn.prepareStatement(sql);
4      st.setInt(0,1);
5      st.execute();
6      java.sql.ResultSet rs =  st.getResultSet();
7      while(rs.next()){
8          String result = rs.getString(colname);
9      }
10   <mapper namespace="org.mybatis.example.BlogMapper">
11           <select id="selectBlog" parameterType="int" resultType="Blog">
12                   select * from Blog where id = #{id}
13           </select>
14   </mapper>
15
```

图 1　JDBC 示例代码

MyBatis 就是将上面这几行代码分解包装。前两行是对数据库的数据源的管理，其中也包括事务管理，3、4 两行 MyBatis 通过配置文件来管理 SQL 以及输入参数的映射，6、7、8 行 MyBatis 获取返回结果到 Java 对象的映射，也是通过配置文件管理。

1.5 MyBatis 和 Hibernate 的对比

MyBatis 与 Hibernate 的对比内容如图 2 所示。

MyBatis	Hibernate
1、是一个SQL语句映射的框架（工具）	1、主流的ORM框架、提供了从 POJO 到数据库表的全套映射机制
2、注重POJO与SQL之间的映射关系。不会为程序员在运行期自动生成 SQL	2、会自动生成全套SQL语句
3、自动化程度低、手工映射SQL，灵活程度高	3、因为自动化程度高、映射配置复杂，api也相对复杂，灵活性低
4、需要开发人员熟炼掌握SQL语句	4、开发人同不必关注SQL底层语句开发

图 2　MyBatis 与 Hibernate 对比

1.6 MyBatis 工作流程

MyBatis 的工作流程如下。

1. 加载配置并初始化

触发条件：加载配置文件。

配置来源于两个地方，一处是配置文件，一处是 Java 代码的注解，将 SQL 的配置信息加载成为一个个 MappedStatement 对象（包括了传入参数映射配置、执行的 SQL 语句、结果映射配置），存储在内存中。

2. 接收调用请求

触发条件：调用 MyBatis 提供的 API。

传入参数：为 SQL 的 ID 和传入参数对象。

处理过程：将请求传递给下层的请求处理层进行处理。

3. 处理操作请求

触发条件：API 接口层传递请求过来。

传入参数：为 SQL 的 ID 和传入参数对象。

处理过程：

①根据 SQL 的 ID 查找对应的 MappedStatement 对象。

②根据传入参数对象解析 MappedStatement 对象，得到最终要执行的 SQL 和执行传入参数。

③获取数据库连接，根据得到的最终 SQL 语句和执行传入参数到数据库执行，并得到执行结果。

④根据 MappedStatement 对象中的结果映射配置对得到的执行结果进行转换处理，并得到最终的处理结果。

⑤释放连接资源。

4. 返回处理结果

将最终的处理结果返回。

小结

MyBatis 是一个支持普通 SQL 查询、存储过程和高级映射的优秀持久层框架。MyBatis 消除了几乎所有的 JDBC 代码和参数的手工设置以及对结果集的检索封装。MyBatis 可以使用简单的 XML 或注解用于配置和原始映射，将接口和 Java 的 POJO（Plain Old Java Objects，普通的 Java 对象）映射成数据库中的记录。

因为 iBatis 已经改名为 MyBatis，后文统称 MyBatis，MyBatis 和 hibernate 一样，是一个 ORM 框架，对我们的数据库操作进行了封装，提高了开发效率。

通过学习可以了解到 MyBatis 只是一个半自动化的 ORM 实现，需要我们自己写 SQL，而不像 hibernate 那样，直接定义好实体与数据表的映射就行。

经典面试题

（1）什么是 MyBatis？
（2）说明 hibarnate 和 ByBatis 的区别。
（3）描述 eclipse 怎么搭建 MyBatis。
（4）简述 MyBatis 的工作流程。
（5）简述 MyBatis 的优点。

第 2 章　MyBatis 基本配置

本章要点（学会后请在方框中打钩）：

- □ 掌握 MyBatis 基本要素
- □ 掌握 configuration.xml
- □ 掌握 MyBatis 环境配置
- □ 掌握事务管理
- □ 掌握数据源配置方法
- □ 掌握 SQL 映射文件

2.1 MyBatis 基本要素

MyBatis 的基本要素如图 3 所示。

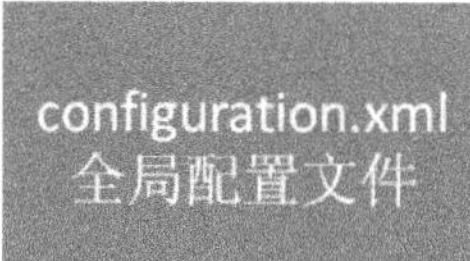

mapper.xml 核心
映射文件

图 3 MyBatis 基本要素

2.2 MyBatis 基础配置文件

MyBatis 的配置文件 configuration.xml 包含了影响 MyBatis 行为甚深的设置(settings)和属性(properties)信息。文档的顶层结构如下：

- configuration 配置
- properties 属性
- settings 设置
- typeAliases 类型命名
- typeHandlers 类型处理器
- objectFactory 对象工厂
- plugins 插件
- environments 环境
- databaseIdProvider 数据库厂商标识
- environment 环境变量
- transactionManager 事务管理器
- dataSource 数据源
- mappers 映射器

2.2.1 基础配置：configuration.xml

```
<!DOCTYPE configuration PUBLIC "-//mybatis.org//DTD Config 3.0//EN" "http://mybatis.org/dtd/mybatis-3-config.dtd">
<configuration>
<environments default="development">
<environment id="development">
<transactionManager type="JDBC"/>
<dataSource type="POOLED">
```

```
<property name="driver" value="${driver}"/>
<property name="url" value="${url}"/>
<property name="username" value="${username}"/>
<property name="password" value="${password}"/>
</dataSource>
</environment>
<environment id="development2">
……
</environment>
</environments>
</configuration>
```

2.2.2 事务管理类型

事务管理类型有如下几种。

（1）JDBC。这个类型直接全部使用 JDBC 的提交和回滚功能，它依靠使用连接的数据源来管理事务的作用域。

（2）MANAGED。这个类型什么都不做，它从不提交、回滚和关闭连接，而是让窗口来管理事务的全部生命周期（比如说 Spring 或者 Java EE 服务器）。

2.2.3 种数据源类型

（1）UNPOOLED。这个数据源的实现只是在每次请求的时候简单地打开和关闭一个连接。虽然有点慢，但作为一些不需要性能和立即响应的简单应用来说，不失为一种好选择。

（2）POOLED。这个数据源的缓存 JDBC 连接对象用于避免每次都要连接和生成连接实例而需要的验证时间 。对于并发 WEB 应用，这种方式非常流行，因为它有最快的响应时间。

（3）JNDI。这个数据源的实现是为了准备和 Spring 或应用服务一起使用，可以在外部也可以在内部配置这个数据源，然后在 JNDI 上下文中引用它。这个数据源配置只需要两个属性。

2.2.4 SQL 映射文件

（1）使用相对路径。

```
<mappers>
    <mapper resource="org/mybatis/builder/UserMapper.xml"/>
    <mapper resource="org/mybatis/builder/AuthorMapper.xml"/>
    <mapper resource="org/mybatis/builder/BlogMapper.xml"/>
```

```
5.         <mapper resource="org/mybatis/builder/PostMapper.xml"/>
6.     </mappers>
```

（2）使用全路径。

```
1.  <mappers>
2.         <mapper url="file:///var/sqlmaps/AuthorMapper.xml"/>
3.         <mapper url="file:///var/sqlmaps/BlogMapper.xml"/>
4.         <mapper url="file:///var/sqlmaps/PostMapper.xml"/>
5.  </mappers>
```

2.3 MyBatis 初体验：CRUD

2.3.1 数据库准备

（1）快速创建一个 Mybatis 数据库，命名为 db_mybatis。

（2）创建一张 user 表，具体结构如图 4 所示，表的数据如图 5 所示。

名	类型	参数1	参数2	不是 nu	
uname	varchar	20	0	☑	1
password	varchar	50	0	☑	

图 4 user 表

uname	password
admin	123456
user1	000000
user2	111111
user3	222222

图 5 user 表的数据

2.3.2 使用 Maven 创建 mybatis 工程

（1）创建 mybatis1_1_ 增删改查，工程结构如图 6 所示。

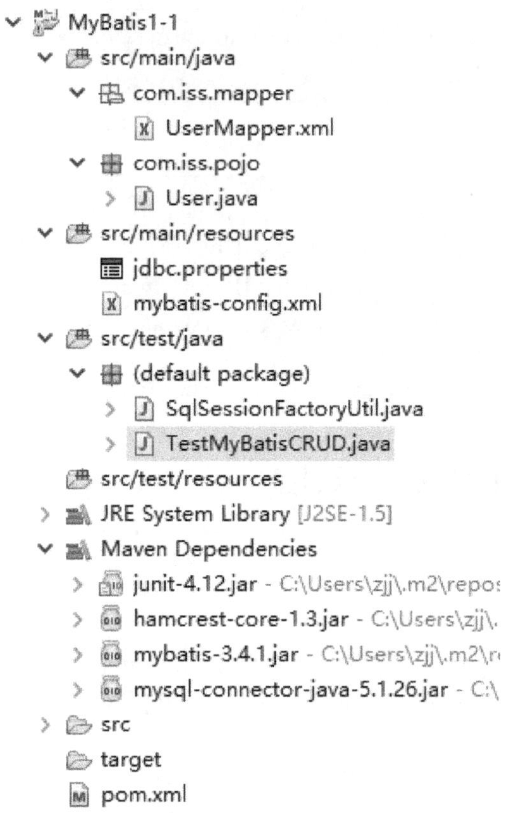

```
MyBatis1-1
  src/main/java
    com.iss.mapper
      UserMapper.xml
    com.iss.pojo
      User.java
  src/main/resources
    jdbc.properties
    mybatis-config.xml
  src/test/java
    (default package)
      SqlSessionFactoryUtil.java
      TestMyBatisCRUD.java
  src/test/resources
  JRE System Library [J2SE-1.5]
  Maven Dependencies
    junit-4.12.jar - C:\Users\zjj\.m2\repo:
    hamcrest-core-1.3.jar - C:\Users\zjj\.
    mybatis-3.4.1.jar - C:\Users\zjj\.m2\r
    mysql-connector-java-5.1.26.jar - C:\
  src
  target
  pom.xml
```

图6　工程结构图

（2）修改 pom.xml 文件，具体内容如下。

```
<project xmlns="http://maven.apache.org/POM/4.0.0"
xmlns:xsi="http://www.w3.org/2001/XMLSchema-instance"
xsi:schemaLocation="http://maven.apache.org/POM/4.0.0
http://maven.apache.org/xsd/maven-4.0.0.xsd">
    <modelVersion>4.0.0</modelVersion>
    <groupId>com.iss.mybatis</groupId>
    <artifactId>MyBatis1-1</artifactId>
    <version>0.0.1-SNAPSHOT</version>
    <dependencies>
      <dependency>
          <groupId>junit</groupId>
          <artifactId>junit</artifactId>
          <version>4.12</version>
      </dependency>
      <dependency>
          <groupId>org.mybatis</groupId>
```

```
                <artifactId>mybatis</artifactId>
                <version>3.4.1</version>
            </dependency>
            <dependency>
                <groupId>mysql</groupId>
                <artifactId>mysql-connector-java</artifactId>
                <version>5.1.26</version>
            </dependency>
        </dependencies>
</project>
```

(3)创建数据库连接文件 jdbc.properties,具体内容如下。

```
jdbc.driverClassName=com.mysql.jdbc.Driver
jdbc.url=jdbc:mysql://localhost:3306/db_mybatis
jdbc.username=root
jdbc.password=root // 数据库密码
```

(4)创建 mybatis-config.xml,具体内容如下。

```
<?xml version="1.0" encoding="UTF-8" ?>
<!DOCTYPE configuration
PUBLIC "-//mybatis.org//DTD Config 3.0//EN"
"http://mybatis.org/dtd/mybatis-3-config.dtd">
<configuration>
    <properties resource="jdbc.properties" />
    <typeAliases>
        <typeAlias alias="user" type="com.iss.pojo.User" />
    </typeAliases>
    <environments default="development">
        <environment id="development">
            <transactionManager type="JDBC" />
            <dataSource type="POOLED">
                <property name="driver" value="${jdbc.driverClassName}" />
                <property name="url" value="${jdbc.url}" />
                <property name="username" value="${jdbc.username}" />
                <property name="password" value="${jdbc.password}" />
            </dataSource>
        </environment>
```

```
        </environments>
        <mappers>
            <mapper resource="com/iss/mapper/UserMapper.xml" />
        </mappers>
</configuration>
```

专家讲解

1）<typeAliases>：表示别名定义，即我们在程序中使用 alias 的定义，即可代表 type 中对应的实体对象。具体见下文中 mapper 文件的使用方法。

2）<mappers>：配置我们数据库语句文件的存放位置，这里只用到了一个文件，因此只配置了一个，更多用法将在后文中进行介绍。

（5）创建 SqlSessionFactoryUtil 文件，用来连接数据库，具体内容如下。

```
public class SqlSessionFactoryUtil {
        private static SqlSessionFactory sqlSessionFactory;
        public static SqlSessionFactory getSqlSessionFactory( ) {
            if (sqlSessionFactory = null) {
                InputStream inputStream = null;
                try {
                inputStream = Resources.getResourceAsStream("mybatis-config.xml");
                sqlSessionFactory = new SqlSessionFactoryBuilder( ).build(inputStream);
                } catch (Exception e) {
                    e.printStackTrace( );
                }
            }
            return sqlSessionFactory;
        }
        public static SqlSession openSession(boolean commit) {
            return getSqlSessionFactory( ).openSession(commit);
        }

}
```

（6）创建 User 实体类，具体内容如下。

```
package com.iss.pojo;
public class User {
        private String uname;
```

```
        private String password;
// 省略 setter 和 getter
        public User(String uname, String password) {
                this.uname = uname;
                this.password = password;
        }
}
```

（7）创建 UserMapper.xml 文件，具体内容如下。

```
<?xml version="1.0" encoding="UTF-8" ?>
        <!DOCTYPE mapper
PUBLIC "-//mybatis.org//DTD Mapper 3.0//EN"
"http://mybatis.org/dtd/mybatis-3-mapper.dtd">
<mapper namespace="com.iss.mapper.userMapper">
        <select id="findUserByUname" parameterType="String" resultType="user">
                select * from user where uname=#{uname}
        </select>
</mapper>
```

专家讲解

（1）namespace。必须与 UserDao 的全路径相匹配，这样才能使的接口与 SQL 文件一一对应。

（2）id。现在必须与接口中的方法名称一致。后续我们将介绍高级用法，使 mapper 文件更加规范化。届时将介绍 id 的另一种实现方式。

（3）parameterType。传入参数类型，对于基本类型，可以使用 Integer、String 等封装类型直接使用。复杂类型，多参数还需要将数据封装成对象才能使用，具体用法见后文中的介绍。

（4）resultType。结果返回值的类型，本例返回的结果是一个 User 对象，这个对象使用了我们前文介绍的别名。如果未使用别名配置，或者上下文中存在多个 User 实体类，最好使用全路径的方式进行配置，防止发生错误。

（8）创建 main 文件，具体内容如下。

```
import org.apache.ibatis.session.SqlSession;
import org.junit.Test;
import com.iss.pojo.User;
public class TestMyBatisCRUD {
        @Test
        public void testRead( ) {
```

```
            SqlSession sqlSession = SqlSessionFactoryUtil.openSession(true);
            String statement = "com.iss.mapper.userMapper.findUserByUname";
            String uname = "admin";
            User user = sqlSession.selectOne(statement, uname);
            if (user != null) {
                System.out.println(" 用户名 :" + user.getUname( ) + ", 密码: " + user.getPassword( ));
            }
        }
}
```

（9）测试运行结果如图 7 所示。

图 7　测试运行结果

2.3.3　修改功能

（1）在 UserMapper.xml 文件中增加 update 语句用来修改用户密码，具体内容如下。

```
<update id="updateUser" parameterType="user">
    update user set password=#{password} where uname=#{uname}
</update>
```

（2）测试修改代码，具体内容如下。

```
@Test
    public void testUpdate( ) {
        SqlSession sqlSession = SqlSessionFactoryUtil.openSession(false);
        String statement = "com.iss.mapper.userMapper.updateUser";
        User user = new User( );
        user.setUname("admin");
        user.setPassword("666666");
        int i = sqlSession.update(statement, user);
        if (i > 0) {
            System.out.println(" 密码修改成功 ");
        }
```

```
            sqlSession.commit( );
            sqlSession.close( );
    }
```

（3）测试运行结果如图 8 所示。

uname	password
admin	666666
user1	000000
user2	111111
user3	222222

密码修改成功

图 8 测试运行结果

专家提示

增删改要通过事务提交。

2.3.4 插入功能

（1）在 UserMapper.xml 文件中增加 insert 语句，具体内容如下。

```
<insert id="insertUser" parameterType="user" >
    insert into user (uname,password) values(#{uname},#{password})
</insert>
```

（2）测试插入功能代码，具体内容如下。

```
@Test
    public void testInsert( ) {
        SqlSession sqlSession = SqlSessionFactoryUtil.openSession(false);
        String statement = "com.iss.mapper.userMapper.insertUser";
        User user = new User( );
        user.setUname("user4");
        user.setPassword("222222");
        int i = sqlSession.update(statement, user);
        if (i > 0) {
            System.out.println(" 插入记录成功 ");
        }
        sqlSession.commit( );
        sqlSession.close( );
    }
```

（3）测试运行结果如图 9。

uname	password
admin	666666
user1	000000
user2	111111
user3	222222
user4	222222

插入记录成功

图 9　测试运行结果

2.4　删除功能

（1）在 UserMapper.xml 文件中增加 delete 语句，具体内容如下。

```
<delete id="deleteUser" parameterType="String">
        delete from user where uname=#{uname}
</delete>
```

（2）测试删除代码。

```
@Test
        public void testDelete( ) {
                SqlSession sqlSession = SqlSessionFactoryUtil.openSession(false);
                String statement = "com.iss.mapper.userMapper.deleteUser";
                String uname = "user4";
                int i = sqlSession.update(statement, uname);
                if (i > 0) {
                        System.out.println(" 删除记录成功 ");
                }
                sqlSession.commit( );
                sqlSession.close( );
        }
```

至此基于 Mave 项目的 MyBatis 配置及其增删改查功能已经完成，如果有任何问题请参看本章项目源码。

小结

本章主要讲述了 MyBatis 基础配置文件的配置方法，重点讲了基于 XML 配置方式的增删改查操作。需要注意的就是增删改操作时一定要使用事务处理，否则不会影响到数据库数据。

经典面试题

(1)MyBatis 的 crud 是什么意思?
(2)怎样新建一个 MyBatis 配置文件?
(3)如何在 MyBatis 的配置文件中引入 properties?
(4)MyBatis 配置文件 Configuration.xml 里的 typeAliases 标签有什么用?
(5)如何让 Intellij IDEA 工具自动创建 MyBatis 配置文件?

跟我上机

建立 Maven 工程,对 flightinfo 表的数据进行 CRUD 操作,表结构如图 10 所示,数据如图 11 所示。

名	类型	长度	小数点	不是 null	
flightid	int	11	0	☑	1
flightnum	varchar	10	0	☑	
flydate	date	0	0	☑	
flytime	varchar	20	0	☑	
startcity	int	11	0	☑	
endcity	int	11	0	☑	
seatnum	int	11	0	☑	

图 10 flightinfo 数据表结构

flightid	flightnum	flydate	flytime	startcity	endcity	seatnum
8	CA1002	2017-07-23	2小时	1	6	50
9	CA1003	2017-08-09	6小时	3	6	44
13	CA1001	2017-08-05	2小时	1	4	50
15	CA1003	2017-08-10	3小时	2	3	15
16	CA1004	2017-08-15	5小时	5	1	250
17	CA1005	2017-08-17	8小时	3	2	50
18	CA1006	2017-08-08	1小时	6	2	50
21	CA1003	2017-08-10	3小时	2	3	150
22	CA1004	2017-08-15	5小时	5	1	250
23	CA1005	2017-08-17	8小时	3	2	50
24	CA1006	2017-08-08	1小时	6	2	50
25	CA1001	2017-08-05	2小时	1	2	50
28	CA1004	2017-08-15	5小时	5	1	250

图 11 flightinfo 表数据信息

第 3 章　configuration.xml 文件配置详解

本章要点(学会后请在方框中打钩):

- □ 了解基础环境配置
- □ 了解事务管理器配置
- □ 了解数据源配置
- □ 了解属性配置
- □ 了解别名设置
- □ 了解映射器设置
- □ 了解 Setting 配置

上一章中，我们演示了最基本的对数据库表进行 CRUD 的操作，下面我们将对上面各项配置文件进行详细的解释，以及正式开发过程中遇到的各种配置的注意事项等。

3.1 基础环境配置：configuration

configuration.xml 基础配置文件代码如下。

```
....
<configuration>
    ......
    <environments default="development">
        <environment id="development">
            <transactionManager type="JDBC" />
            <dataSource type="POOLED">
                <property name="driver" value="${jdbc.driverClassName}" />
                <property name="url" value="${jdbc.url}" />
                <property name="username" value="${jdbc.username}" />
                <property name="password" value="${jdbc.password}" />
            </dataSource>
        </environment>
    </environments>
    ......
</configuration>
```

专家讲解

（1）默认的环境 ID（比如 :default="development"）。在实际使用时，常见的环境配置有开发环境、测试环境、正式环境等。为了方便开发，MyBatis 提供给我们方便切换环境的方法，这里的 ID 需要对应下面的每个 environment 标签的 ID 值。

（2）每个 environment 元素定义的环境 ID（比如 :id="development"）。在该标签中，我们需要配置每一个环境的具体参数，本例只用到了基本配置。

3.2 事务管理器的配置：transactionManager

在 MyBatis 中有以下两种类型的事务管理器（也就是 type="[JDBC|MANAGED]"）。

（1）JDBC。这个配置就是直接使用了 JDBC 的提交和回滚设置，它依赖于从数据源得到的连接来管理事务范围。

（2）MANAGED。这个配置几乎没做什么。它从来不提交或回滚一个连接，而是让容器来管理事务的整个生命周期(比如 JEE 应用服务器的上下文)。默认情况下它会关闭连接,然而一些容器并不希望这样,因此需要将 closeConnection 属性设置为 false 来阻止它默认的关闭行为。

专家讲解

如果正在使用 Spring+MyBatis,则没有必要配置事务管理器,因为 Spring 模块会使用自带的事务管理器来覆盖前面的配置。

3.3 数据源的配置:dataSource

3.3.1 POOLED

这种数据源的实现利用"池"的概念将 JDBC 连接对象组织起来,避免了创建新的连接实例时所必需的初始化和认证时间。这是一种使得并发 Web 应用快速响应请求的流行处理方式。

POOLED 类型的数据源可根据需要配置以下属性。

(1)poolMaximumActiveConnections。在任意时间可以存在的活动(也就是正在使用)连接数量,默认值:10。

(2)poolMaximumIdleConnections。任意时间可能存在的空闲连接数。

(3)poolMaximumCheckoutTime。在被强制返回之前,池中连接被检出(checked out)时间,默认值:20000 毫秒(即 20 秒)。

(4)poolTimeToWait。这是一个底层设置,如果获取连接花费的相当长的时间,它会给连接池打印状态日志并重新尝试获取一个连接(避免在误配置的情况下一直安静的失败),默认值:20000 毫秒(即 20 秒)。

(5)poolPingQuery。发送到数据库的侦测查询,用来检验连接是否处在正常工作秩序中并准备接受请求。默认是"NO PING QUERY SET",这会导致多数数据库驱动失败时带有一个恰当的错误消息。

(6)poolPingEnabled。是否启用侦测查询。若开启,也必须使用一个可执行的 SQL 语句设置 poolPingQuery 属性(最好是一个非常快的 SQL),默认值:false。

(7)poolPingConnectionsNotUsedFor。配置 poolPingQuery 的使用频度。这可以被设置成匹配具体的数据库连接超时时间,来避免不必要的侦测,默认值: 0(即所有连接每一时刻都被侦测;当且仅当 poolPingEnabled 为 true 时适用)。

3.3.2 UNPOOLED

这个数据源的实现只是每次被请求时打开和关闭连接。虽然有一点慢,它对在及时可用连接方面没有性能要求的简单应用程序是一个很好的选择。不同的数据库在这方面的表现也是不一样的,所以对某些数据库来说使用连接池并不重要,这个配置也是理想的。

UNPOOLED 类型的数据源仅仅需要配置以下 5 种属性：

（1）driver。这是 JDBC 驱动的 Java 类的完全限定名（并不是 JDBC 驱动中可能包含的数据源类）。

（2）url。这是数据库的 JDBC URL 地址。

（3）username。登录数据库的用户名。

（4）password。登录数据库的密码。

（5）defaultTransactionIsolationLevel。默认的连接事务隔离级别。

3.3.2 JNDI

这个数据源的实现是为了能在如 EJB 或应用服务器这类容器中使用，容器可以集中或在外部配置数据源，然后放置一个 JNDI 上下文的引用。这种数据源的配置只需要以下两个属性。

（1）initial_context。这个属性用来在 InitialContext 中寻找上下文，即，initialContext.lookup(initial_context)。这是个可选属性，如果忽略，那么 data_source 属性将会直接从 InitialContext 中寻找。

（2）data_source。这是引用数据源实例位置的上下文的路径。提供了 initial_context 配置时会在其返回的上下文中进行查找，没有提供时则直接在 InitialContext 中查找。

3.4 属性配置：properties

属性配置有如下两种。

（1）动态配置，即通过 porpeties 文件来配置。

（2）子元素配置，即通过内置数据来配置，具体代码如下。

```
<!-- <properties resource="jdbc.properties" />
本注释和下面配置功能相同，优先级不同
-->
<properties>
<property name="jdbc.driverClassName" value="com.mysql.jdbc.Driver"/>
<property name="jdbc.url" value="jdbc:mysql://localhost:3306/db_mybatis"/>
<property name="jdbc.username" value="root"/>
<property name="jdbc.password" value="root"/>
</properties>
```

专家讲解

如果属性在不只一个地方进行了配置，那么 MyBatis 将按照下面的顺序来加载。

（1）在 properties 元素体内指定的属性首先被读取。

（2）然后根据 properties 元素中的 resource 属性读取类路径下属性文件或根据 url 属性指定的路径读取属性文件，并覆盖已读取的同名属性。

（3）最后读取作为方法参数传递的属性，并覆盖已读取的同名属性。

因此，通过方法参数传递的属性具有最高优先级，resource/url 属性中指定的配置文件次之，最低优先级的是 properties 属性中指定的属性。

3.5 别名配置：typeAliases

类型别名是为 Java 类型设置一个短的名字，它只和 XML 配置有关，存在的意义仅在于用来减少类完全限定名称的冗余。

别名配置的基本代码如下。

```
<typeAliases>
        <typeAlias alias="user" type="com.iss.pojo.User" />
    </typeAliases>
```

这种用法会使得配置项非常多，因此推荐使用下面的方法进行配置。

```
<typeAliases>
<package name="com.iss.pojo"/>
<!-- <typeAlias alias="user" type="com.iss.pojo.User" /> -->
    </typeAliases>
```

专家讲解

（1）配置一个包路径，当前包下面的所有类都会被自动转化为一个 JAVABEAN。

（2）特别的，当使用这种方法配置时，其别名是首字母小写的别名，如，User 类对象的别名为 user。同时，我们不难发现，在没有其他配置的情况下，最好不要出现相同命名的对象。如果实在需要时，要使用单独注解的方式来实现，具体使用如下：

```
@Alias("user")
public class User {
    private String uname;
    private String password;
}
```

另外，在 MyBatis 中已经创建了一些基本的别名对象，它们都是大小写不敏感的，需要注意的是由基本类型名称重复导致的特殊处理。具体数据如表 1 所示。

表 1 MyBatis 数据类型

别名	映射的类型	别名	映射的类型
_byte	byte	double	Double
_long	long	float	Float
_short	short	boolean	Boolean
_int	int	date	Date
_integer	int	decimal	BigDecimal
_double	double	bigdecimal	BigDecimal
_float	float	object	Object
_boolean	boolean	map	Map
string	String	hashmap	HashMap
byte	Byte	list	List
long	Long	arraylist	ArrayList
short	Short	collection	Collection
int	Integer	iterator	Iterator
integer	Integer		

3.6 映射器配置(mappers)

所谓映射器，即 MyBatis 寻找数据库语句文件的配置。

3.6 1 使用相对路径进行配置

```
<mappers>
    <mapper resource="com/iss/mapper/UserMapper.xml" />
</mappers>
```

3.6.2 使用文件定位符配置

```
<mappers>
    <mapper url="file:///var/mappers/UserMapper.xml"/>
</mappers>
```

3.6.3 使用接口配置

```
<mappers>
        <mapper class="com.iss.mapper.UserMapper"/>
</mappers>
```

3.6.4 使用包路径配置

```
<mappers>
        <package name="com.iss.mapper"/> // 尝试使用注解方式时使用
</mappers>
```

3.7 Setting 配置

这是 MyBatis 中极为重要的配置项，它们的属性值能够改变 MyBatis 的运行状态，请小心配置。

Setting 的具体代码如下，相关解释如表 2 所示。

```
<settings>
    <setting name="cacheEnabled" value="true"/>
    <setting name="lazyLoadingEnabled" value="true"/>
    <setting name="multipleResultSetsEnabled" value="true"/>
    <setting name="useColumnLabel" value="true"/>
    <setting name="useGeneratedKeys" value="false"/>
    <setting name="autoMappingBehavior" value="PARTIAL"/>
    <setting name="autoMappingUnknownColumnBehavior" value="WARNING"/>
    <setting name="defaultExecutorType" value="SIMPLE"/>
    <setting name="defaultStatementTimeout" value="25"/>
    <setting name="defaultFetchSize" value="100"/>
    <setting name="safeRowBoundsEnabled" value="false"/>
    <setting name="mapUnderscoreToCamelCase" value="false"/>
    <setting name="localCacheScope" value="SESSION"/>
    <setting name="jdbcTypeForNull" value="OTHER"/>
    <setting name="lazyLoadTriggerMethods" value="equals,clone,hashCode,toString"/>
</settings>
```

表 2 Settingname 属性相关解释

参数名称	功能描述	有效配置	默认值
cacheEnabled	该配置影响的所有映射器中配置的缓存的全局开关	true|false	true
lazyLoadingEnabled	延迟加载的全局开关。当开启时，所有关联对象都会延迟加载。特定关联关系中可通过设置 fetchType 属性来覆盖该项的开关状态	true|false	true
multipleResultSetsEnabled	是否允许单一语句返回多结果集（需要兼容驱动）	true|false	true
useColumnLabel	使用列标签代替列名。不同的驱动在这方面会有不同的表现，具体可参考相关驱动文档或通过测试这两种不同的模式来观察所用驱动的结果	true|false	true
useGeneratedKeys	允许 JDBC 支持自动生成主键，需要驱动兼容。如果设置为 true 则这个设置强制使用自动生成主键，尽管一些驱动不能兼容但仍可正常工作（比如 Derby）	true|false	False
autoMappingBehavior	指定 MyBatis 应如何自动映射列到字段或属性。NONE 表示取消自动映射；PARTIAL 只会自动映射没有定义嵌套结果集映射的结果集。FULL 会自动映射任意复杂的结果集（无论是否嵌套）	NONE,PARTIAL,FULL	PARTIAL
autoMappingUnknownColumnBehavior	Specifythebehaviorwhendetectsanunknown-column(orunknownpropertytype)ofautomatic-mappingtarget. NONE:Donothing WARNING:Outputwarninglog(Thelog-levelof'org.apache.ibatis.session.AutoMappingUnknownColumnBehavior'mustbeset-toWARN) FAILING:Failmapping(ThrowSqlSessionException)	NONE,WARNING,FAILING	NONE
defaultExecutorType	配置默认的执行器。SIMPLE 就是普通的执行器；REUSE 执行器会重用预处理语句（preparedstatements）；BATCH 执行器将重用语句并执行批量更新	SIMPLEREUSE-BATCH	SIMPLE

续表

参数名称	功能描述	有效配置	默认值
defaultStatement-Timeout	设置超时时间，它决定驱动等待数据库响应的秒数	Anypositiveinteger	NotSet(null)
defaultFetchSize	Setsthedriverahintastocontrolfetchingsizeforre-turnresults.Thisparametervaluecanbeoverrideb-yaquerysetting.	Anypositiveinteger	NotSet(null)
safeRowBound-sEnabled	允许在嵌套语句中使用分页（RowBounds）	true|false	False
mapUnderscore-ToCamelCase	是否开启自动驼峰命名规则（camelcase）映射，即从经典数据库列名 A_COLUMN 到经典 Java 属性名 aColumn 的类似映射	true|false	False
localCacheScope	MyBatis 利用本地缓存机制（LocalCache）防止循环引用（circularreferences）和加速重复嵌套查询。默认值为 SESSION，这种情况下会缓存一个会话中执行的所有查询。若设置值为 STATEMENT，本地会话仅用在语句执行上，对相同 SqlSession 的不同调用将不会共享数据	JdbcTypeenumera-tion.Mostcom-monare:NULL,-VARCHARandO-THER	OTHER
jdbcTypeForNull	当没有为参数提供特定的 JDBC 类型时，为空值指定 JDBC 类型。某些驱动需要指定列的 JDBC 类型，多数情况直接用一般类型即可，比如 NULL、VARCHAR 或 OTHER	JdbcTypeenumera-tion.Mostcom-monare:NULL,-VARCHARandO-THER	OTHER
lazyLoadTrigger-Methods	指定哪个对象的方法触发一次延迟加载	Amethodnamelist-separatedbycom-mas	equals,clone,hash-Code,toString

3.8 typeHandlers 配置

无论是 MyBatis 在预处理语句（PreparedStatement）中设置一个参数时，还是从结果集中取出一个值时，都会用类型处理器将获取的值以合适的方式转换成 Java 类型。表 3 中描述了一些默认的类型处理器。

表 3 默认的类型处理器

类型处理器	Java 类型	JDBC 类型
BooleanTypeHandler	java.lang.Boolean,boolean	数据库兼容的 BOOLEAN

续表

类型处理器	Java 类型	JDBC 类型
ByteTypeHandler	java.lang.Byte,byte	数据库兼容的 NUMERIC 或 BYTE
ShortTypeHandler	java.lang.Short,short	数据库兼容的 NUMERIC 或 SHORTINTEGER
IntegerTypeHandler	java.lang.Integer,int	数据库兼容的 NUMERIC 或 INTEGER
LongTypeHandler	java.lang.Long,long	数据库兼容的 NUMERIC 或 LONGINTEGER
FloatTypeHandler	java.lang.Float,float	数据库兼容的 NUMERIC 或 FLOAT
DoubleTypeHandler	java.lang.Double,double	数据库兼容的 NUMERIC 或 DOUBLE
BigDecimalTypeHandler	java.math.BigDecimal	数据库兼容的 NUMERIC 或 DECIMAL
StringTypeHandler	java.lang.String	CHAR,VARCHAR
ClobReaderTypeHandler	java.io.Reader	—
ClobTypeHandler	java.lang.String	CLOB,LONGVARCHAR
NStringTypeHandler	java.lang.String	NVARCHAR,NCHAR
NClobTypeHandler	java.lang.String	NCLOB
BlobInputStreamTypeHandler	java.io.InputStream	—
ByteArrayTypeHandler	byte[]	数据库兼容的字节流类型
BlobTypeHandler	byte[]	BLOB,LONGVARBINARY
DateTypeHandler	java.util.Date	TIMESTAMP
DateOnlyTypeHandler	java.util.Date	DATE
TimeOnlyTypeHandler	java.util.Date	TIME
SqlTimestampTypeHandler	java.sql.Timestamp	TIMESTAMP
SqlDateTypeHandler	java.sql.Date	DATE
SqlTimeTypeHandler	java.sql.Time	TIME
ObjectTypeHandler	Any	OTHER 或未指定类型
EnumTypeHandler	EnumerationType	VARCHAR- 任何兼容的字符串类型，存储枚举的名称（而不是索引）
EnumOrdinalTypeHandler	EnumerationType	任何兼容的 NUMERIC 或 DOUBLE 类型，存储枚举的索引（而不是名称）

小结

本章主要讲解了基础配置文件 configuration.xml 的结构和配置细节，详细介绍了数据源、属性、别名、映射器等。让大家参考并熟悉 Setting 的设计和 typeHandlers 配置进行数据转换是所使用的类型处理器。

经典面试题

（1）MyBatis 入门配置都需要哪些包？
（2）用 MyBatis SQL 语句都在配置文件里写吗？
（3）MyBatis 主要的配置文件有哪些？
（4）MyBatis 怎么配置使用多个数据源？
（5）MyBatis 中如何映射实体类和表名？

跟我上机

新建一个基于 Maven 的 Java 项目完成一个给指定的员工进行投票的功能。
要求：①读出所有员工，显示当前票数。
②能够给指定员工进行投票。
③数据库表自行设计。

第 4 章　MyBatis 映射文件配置详解

本章要点（学会后请在方框中打钩）：

- ☐ 采用映射文件的方式对数据库进行增删改查配置
- ☐ 掌握 ResultMaps 的用法
- ☐ 掌握嵌套查询和嵌套结果的用法与配置
- ☐ 掌握结果集的关联配置

4.1 映射文件

映射文件即 MyBatis 将 SQL 语句与程序接口进行映射匹配的文件，在实际开发中是一一对应的关系。

MyBatis 真正的强大之处，在于它对 SQL 的映射，这也正是它吸引人的地方。实现相同的功能，它要比直接使用 JDBC 省去 95% 的代码量。而且将 SQL 语句独立在 Java 代码之外，为程序的修改和纠错提供了更大的灵活性，可以直接修改 SQL 语句，而无须重新编译 Java 程序。

SQL 映射文件也是 XML 格式，其顶级元素有以下几个。

（1）SELECT——映射 SQL 查询语句。

（2）INSERT——映射 SQL 插入语句。

（3）UPDATE——映射 SQL 更新语句。

（4）DELETE——映射 SQL 删除语句。

（5）SQL——就像程序中可以复用的函数一样，这个元素下放置可以被其他语句重复引用的 SQL 语句。

（6）RESULTMAP——用来描述如何从数据库查询结果集中来加载对象。

（7）CACHE——给定命名空间的缓存配置。

（8）CACHE-REF——其他命名空间缓存配置引用。

4.1.1 查询语句详解

查询语句是 SQL 中使用频率最高的语句，MyBatis 中的查询语句映射也非常简单，示例代码如下。

```
<select id="findUserById" parameterType="String" resultType="User">
        select * from sysuser where id=#{id}
</select>
```

专家讲解

（1）id：必须要与 dao 接口中的方法名称一一对应。

（2）parameterType：传入参数的类型，Java 基本类型可以直接书写。

（3）resultType：结果类型，这里我们返回的一个具体的 Java 对象

（4）#{id}：取出传入参数的值。括号内的值需要与传入的形参保持一致。

这样配置之后，其运行效果等同于如下 JDBC 语句。

```
String selectPerson = "SELECT * FROM PERSON WHERE ID=?";
PreparedStatement ps = conn.prepareStatement(selectPerson);
ps.setInt(1,id);
```

SELECT 的其他可用完整参数配置如下所示。

```
<select id="selectPerson" parameterType="int" parameterMap="deprecated"
resultType="hashmap" resultMap="personResultMap" flushCache="false" useCache="true"
timeout="10000" fetchSize="256" statementType="PREPARED"
resultSetType="FORWARD_ONLY">
```

SELECT 属性解释如表 1 所示。

表 1 SELECT 属性解释参考表

属性	描述
id	命名空间中唯一的标识符,可以被用来引用这条语句
parameterType	将会传入这条语句的参数类的完全限定名或别名。这个属性是可选的,因为 MyBatis 可以通过 TypeHandler 推断出具体传入语句的参数,默认值为 unset
resultType	从这条语句中返回的期望类型的类的完全限定名或别名。注意如果是集合情形,那应该是集合可以包含的类型,而不能是集合本身。使用 resultType 或 resultMap,但不能同时使用
resultMap	外部 resultMap 的命名引用。结果集的映射是 MyBatis 最强大的特性,对其有一个很好的理解的话,许多复杂映射的情形都能迎刃而解。使用 resultMap 或 resultType,但不能同时使用
flushCache	将其设置为 true,任何时候只要语句被调用,都会导致本地缓存和二级缓存被清空,默认值:false
useCache	将其设置为 true,将会导致本条语句的结果被二级缓存,默认值:对 select 元素为 true
timeout	这个设置是在抛出异常之前,驱动程序等待数据库返回请求结果的秒数。默认值为 unset(依赖驱动)
fetchSize	这是尝试影响驱动程序每次批量返回的结果行数和这个设置值相等。默认值为 unset(依赖驱动)
statementType	STATEMENT, PREPARED 或 CALLABLE 的一个。这会让 MyBatis 分别使用 Statement,PreparedStatement 或 CallableStatement,默认值:PREPARED
resultSetType	FORWARD_ONLY,SCROLL_SENSITIVE 或 SCROLL_INSENSITIVE 中的一个,默认值为 unset(依赖驱动)
databaseId	如果配置了 databaseIdProvider,MyBatis 会加载所有的不带 databaseId 或匹配当前 databaseId 的语句;如果带或者不带的语句都有,则不带的会被忽略
resultOrdered	这个设置仅针对嵌套结果 select 语句适用:如果为 true,就是假设包含了嵌套结果集或是分组,这样的话当返回一个主结果行的时候,就不会发生有对前面结果集的引用的情况。这就使得在获取嵌套的结果集的时候不至于导致内存不够用。默认值:false
resultSets	这个设置仅对多结果集的情况适用,它将列出语句执行后返回的所有结果集并为其中每个结果集起一个名称,名称是逗号分隔的

4.1.2 插入、更新、删除配置详解

插入、更新、删除语句和查询语句类似，也需要定义唯一的 id，指定传入参数的类型。数据的更新语句（增删改）配置十分类似，具体代码如下。

```
// 插入语句
<insert
    id="insertAuthor"
    parameterType="domain.blog.Author"
    flushCache="true"
    statementType="PREPARED"
    keyProperty=""
    keyColumn=""
    useGeneratedKeys=""
    timeout="20">
```

```
 // 更新语句
<update
  id="updateAuthor"
  parameterType="domain.blog.Author"
  flushCache="true"
  statementType="PREPARED"
  timeout="20">
```

```
// 删除语句
<delete
    id="deleteAuthor"
    parameterType="domain.blog.Author"
    flushCache="true"
    statementType="PREPARED"
    timeout="20">
```

表 2　插入、更新、删除属性表

属性	描述
useGeneratedKeys	（仅对 insert 和 update 有用）这会令 MyBatis 使用 JDBC 的 getGeneratedKeys 方法来取出由数据库内部生成的主键（比如：像 MySQL 和 SQLServer 这样的关系数据库管理系统的自动递增字段），默认值：false

续表

属性	描述
keyProperty	（仅对 insert 和 update 有用）唯一标记一个属性，MyBatis 会通过 getGeneratedKeys 的返回值或者通过 insert 语句的 selectKey 子元素设置它的键值，默认：unset。如果希望得到多个生成的列，也可以是逗号分隔的属性名称列表
keyColumn	（仅对 insert 和 update 有用）通过生成的键值设置表中的列名，这个设置仅在某些数据库（像 PostgreSQL）是必须的，当主键列不是表中的第一列的时候需要设置。如果希望得到多个生成的列，也可以是逗号分隔的属性名称列表
databaseId	如果配置了 databaseIdProvider，MyBatis 会加载所有的不带 databaseId 或匹配当前 databaseId 的语句；如果带或者不带的语句都有，则不带的会被忽略

从上面的解释可以看出，插入语句拥有更多的配置项可供选择，而数据更改语句则相对少一些。下面针对新增语句做进一步的说明。

1）单行插入时，主键自动生成

此时需要设置 useGeneratedKeys="true"，然后再把 keyProperty 设置到目标属性上，具体代码如下。

```
<insert id="insertAuthor" useGeneratedKeys="true" keyProperty="id">
  insert into Author (username,password,email,bio)
  values (#{username},#{password},#{email},#{bio})
</insert>
```

2）多行插入时，主键自动生成

关于 <foreach> 请关注后面的章节。

```
<insert id="insertAuthor" useGeneratedKeys="true"
    keyProperty="id">
  insert into Author (username, password, email, bio) values
  <foreach item="item" collection="list" separator=",">
    (#{item.username}, #{item.password}, #{item.email}, #{item.bio})
  </foreach>
  </insert>
```

4.2 resultMap 基本用法

resultMap 能够将我们查询数据的结果集合在书写配置时一劳永逸。并且，其能够支持复杂的数据类型。

4.2.1 resultMap 的基本用法

（1）首先来看看基本的映射语句，代码内容如下。

```
<select id="findUserByUname" resultType="userResultMap">
    select uname,password  from user  where uname = #{uname}
</select>
```

（2）对应的属性 resultType=" userResultMap"，需要在 mapper 文件中加入如下内容。

```
<resultMap id="userResultMap" type="User">
    <id property="uname" column="uname" />
    <result property="password" column="password"/>
</resultMap>
```

专家讲解

（1）外层“id”属性：即 SQL 语句用引用的标识，其值需要在该文件中具有唯一性。

（2）外层“type”属性：即结果返回的目标对象（user），这里需要定义为一个全路径标识的对象，或者使用别名的对象。

（3）内层“id”标签：即当前结果集的主键。

（4）内层“property”属性：即当前结果返回的目标对象的属性（user 中的属性）。

（5）内层“column”属性：即在数据库中对应的列名。

除此之外，内层可以使用的属性还有以下几点。

（1）“javaType”：一个 Java 类的完全限定名，或一个类型别名。如果你映射到一个 JavaBean,MyBatis 通常可以断定类型。如果你映射到的是 HashMap, 那么你应该明确地指定 javaType 来保证所需的行为。

（2）“jdbcType”：下面的表格列出了所支持的 JDBC 类型。JDBC 类型是仅仅需要对插入、更新和删除操作可能为空的列进行处理。这是 JDBCjdbcType 的需要，而不是 MyBatis 的。

（3）“typeHandler”：默认的类型处理器。使用这个属性，可以覆盖默认的类型处理器。这个属性值是类的完全限定名或者是一个类型处理器的实现，或者是类型别名。

（4）MyBatis 支持以下 JDBC 类型。

BIT	FLOAT	CHAR	TIMESTAMP	OTHER	UNDEFINED
TINYINT	REAL	VARCHAR	BINARY	BLOG	NVARCHAR
SMALLINT	DOUBLE	LONGVARCHAR	VARBINARY	CLOB	NCHAR
INTEGER	NUMERIC	DATE	LONGVARBINARY	BOOLEAN	NCLOB
BIGINT	DECIMAL	TIME	NULL	CURSOR	ARRAY

(3) 解决数据库中的列名与 java 对象的属性名不一致，具体代码内容如下。

```
<select id="selectUsers" resultType="User">
  select
    user_id  as "id",  //user_id 是数据库表的字段，id 是 User 实体类的属性
    user_password  as "password"
  from sysuser
  where id = #{id}
</select>
```

4.2.2 resultMap 的高级用法：构造方法

现在已经掌握了单表查询，但是我们日常的业务中也存在着多表关联查询，结果是复杂的数据集合等情况。下面来介绍一下 ResultMaps 的高级用法。

resultMap 标记中存在很多的子元素，下面逐一进行介绍。

(1)“constructor”：类在实例化时，用来注入结果到构造方法中。

(2)“idArg”：ID 参数，标记结果作为 ID，可以帮助提高整体的效率。

(3)“arg”：注入到构造方法的一个不同结果。

(4)“id”：类似于数据库的主键，能够帮助提高整体的效率。

(5)“result”：即结果字段，其中包括 Java 对象的属性值和数据库列名。

(6)“association”：复杂类型的结果关联，结果映射能够关联自身或者另一个结果集。

(7)“collection”：复杂类型的集合，结果映射自身或者映射结果集。

(8)“discriminator”：使用结果值来决定使用哪个结果映射。

(9)“case”：基于某些值的结果映射。嵌入结果映射，这种情形也映射到它本身，因此能够包含相同的元素或者参照一个外部的结果映射。

下面开始详细说明每一个元素，请一定按照单元测试的方法推进，千万不要一次性配置大量属性，以免打消学习兴趣。

```
<select id="getStudent" resultMap="getStudentRM">
  SELECT ID, Name, Age
    FROM TStudent
</select>
<resultMap id="getStudentRM" type="EStudnet">
  <constructor>
    <idArg column="ID" javaType="_long"/>
    <arg column="Name" javaType="String"/>
    <arg column="Age" javaType="_int"/>
  </constructor>
</resultMap>
```

专家讲解

在 MyBatis 中，为了向这个构造方法中注入结果，MyBatis 需要通过它的参数类型来表示构造方法。Java 中，没有反射参数名称的方法，因此，当创建一个构造方法的元素时，必须保证参数是按照顺序排列的，而且，数据类型也必须匹配。

4.2.3 resultMap 的高级用法：关联查询

MyBatis 可以很方便地把 SQL 选择出来的数据直接映射为对象的属性，把对象取出来。但是有些对象的属性是集合类型，集合里保存的是数个其他类型的对象。如何用 MyBatis 把它取出来呢？这时就需要用关联查询了。

关联查询的不同之处是我们必须告诉 MyBatis 如何加载关联关系，这里有以下两种方法。

（1）嵌套查询：即通过执行另一个预期返回复杂类型的 SQL 语句。

（2）嵌套结果：使用嵌套结果映射来处理联合结果中重复的子集。

在正式使用之前先来看看这个属性配置的具体含义，这里还需要注意属性配置与前面的增改删查的区别。关联查询属性如表 3 所示。

表 3　关联查询属性

属性	描述
property	映射到列结果的字段或属性。如果匹配的是存在的，和给定名称相同的 property-JavaBeans 的属性，那么就会使用。否则 MyBatis 将会寻找给定名称的字段。这两种情形你可以使用通常点式的复杂属性导航。比如，可以这样映射："username"，或者映射到一些复杂的属性："address.street.number"
javaType	一个 Java 类的完全限定名，或一个类型别名（参考上面内建类型别名的列表）。如果你映射到一个 JavaBean,MyBatis 通常可以断定类型。然而，如果你映射到的是 HashMap, 那么你应该明确地指定 javaType 来保证所需的行为
jdbcType	在这个表格之前的所支持的 JDBC 类型列表中的类型。JDBC 类型是仅仅需要对插入、更新和删除操作可能为空的列进行处理。这是 JDBC 的需要 ,jdbcType 而不是 MyBatis 的。如果你直接使用 JDBC 编程，你需要指定这个类型，但仅仅对可能为空的值
typeHandler	我们在前面讨论过默认的类型处理器。使用这个属性，你可以覆盖默认的 typeHandler 类型处理器。这个属性值是类的完全限定名或者是一个类型处理器的实现，或者是类型别名

现在正式介绍这两种方式。

1）嵌套查询

嵌套查询属性如表 4 所示。

表 4　嵌套查询属性

属性	描述
column	这是来自数据库的类名，或重命名的列标签的值作为一个输入参数传递给嵌套语句，这和通常传递给 resultSet.getString(columnName) 方法的字符串是相同的。 注意：要处理复合主键，你可以指定多个列名通过 column="{prop1=col1,prop2=col2}" 这种语法来传递给嵌套查询语句。这会引起 prop1 和 prop2 以参数对象形式来设置给目标嵌套查询语句
select	另外一个映射语句的 ID，将会按照属性的映射来加载复杂类型。获取的在列属性中指定的列的值将被传递给目标 select 语句作为参数。表格后面有一个详细的示例。 注意：要处理复合主键，可以通过使用 column="{prop1=col1,prop2=col2}" 这种语法指定多个列名传递给嵌套查询语句。这会导致 prop1 和 prop2 以参数对象形式来设置给目标嵌套查询语句
fetchType	可选，有效值包括 lazy 和 eager。如果存在，将在当前映射关系中取代全局变量 lazyLoading-Enabled

嵌套查询实例演示代码如下。

```
<resultMap id="blogResult" type="Blog">
  <association property="author" column="author_id" javaType="Author" select="selectAuthor"/>
</resultMap>

<select id="selectBlog" resultMap="blogResult">
  SELECT * FROM BLOG WHERE ID = #{id}
</select>

<select id="selectAuthor" resultType="Author">
  SELECT * FROM AUTHOR WHERE ID = #{id}
</select>
```

2）嵌套结果

首先来看看有哪些属性可以使用，具体内容见表 5。

表 5　嵌套结果属性

属性	描述
resultMap	这是结果映射的 ID, 可以映射关联的嵌套结果到一个合适的对象图中。这是一种替代方法来调用另外一个查询语句。这允许你联合多个表来合成到 resultMap 一个单独的结果集。这样的结果集可能包含需要被分解的相同的,重复的数据组并且合理映射到一个嵌套的对象图。为了使它变得容易 ,MyBatis 让你"链接"结果映射 , 来处理嵌套结果。下面给予一个很容易来仿照例子
columnPrefix	当连接多个表时,你最好使用列的别名来避免在一个结果集合中出现的名称重复。对于制定的前缀,MyBatis 允许我们映射列到外部集合中,具体用法请参照后面的例子
notNullColumn	只有在至少有一个非空列映射到子对象的属性时,才创建一个默认的子对象。通过这个属性,我们可以设置哪一个列必须有值来改变这个行为,此时的 MyBatis 就会按照这个非空设置来创建一个子对象。多个列存在时,可以通过逗号作为分割符。默认情况下,该属性是不会被设置的,即 unset
autoMapping	如果存在此属性的话, MyBatis 会在映射到对应属性时启用或者禁用自动映射的功能。这个属性将会在全局范围内覆盖自动映射的功能。 注意:该属性没有对外部结果集造成影响。因此,在 select 或者结果集合中使用是没有意义的。默认情况下,它是不设置的,即 unset

嵌套结果实例演示代码内容如下,用以实现销售与客户多对多关系。

```
<resultMap id="salesResultMap" type="com.iss.pojo.Sales">
    <id property="salesId" column="sales_id" />
    <result property="salesName" column="sales_name" />
    <result property="phone" column="sales_phone" />
    <result property="fax" column="sales_fax" />
    <result property="email" column="sales_email" />
    <result property="isValid" column="is_valid" />
    <result property="createdTime" column="created_time" />
    <result property="updateTime" column="update_time" />

    <!-- 定义多对一关联信息 ( 嵌套查询方式 ) -->
    <!-- <association property="userInfo" column="user_id" javaType="User"
        select="selectUser" fetchType="lazy"> </association> -->

    <!-- 定义多对一关联信息 ( 嵌套结果方式 ) -->
    <association property="userInfo" resultMap="com.iss.mapper.userResult" />

    <!-- 定义一对多集合信息 ( 每个销售人员对应多个客户 ) -->
```

```
<!-- <collection property="customers" column="sales_id" select="getCustomerForSales"  /> -->

<collection property="customers" ofType="com.iss.pojo.Customer">
    <id property="customerId" column="customer_id" />
    <result property="customerName" column="customer_name" />
    <result property="isValid" column="is_valid" />
    <result property="createdTime" column="created_time" />
    <result property="updateTime" column="update_time" />
        <association property="userInfo" resultMap="com.iss.mapper.userResult" columnPrefix="cu_" />
</collection>
</resultMap>
```

有能力的同学可自行完成上述实例。

4.3 综合实例演示

以一个教师对应多个课程为例，进行如下代码演练。

4.3.1 编写实体类

```
public class Course{// 课程类
    int id;
    String name;
}

public class Tutor{// 教师类
    int id;
    String name;
    List<Course> courses;
}
```

4.3.2 写出两者的 resultMap

```
<resultMap type="Course" id="courseResult">
        <result column="course_id" property="id" />
        <result column="course_name" property="name" />
```

```
    </resultMap>

    <resultMap type="Tutor" id="tutorResult">
        <id column="tutor_id" property="id" />
        <result column="tutor_name" property="name" />
        <collection property="courses" resultMap="Course" />
</resultMap>
```

4.3.3 select 语句的 resultMap 设为 tutorResult

```
<select id="findTutorById" parameterType="int" resultMap="TutorResult">
    SELECT TUTOR_ID, TUTOR_NAME, COURSE_ID, COURSE_NAME FROM TUTOR
</select>
```

4.3.4 测试结果

测试结果如表 6 所示。

表 6 测试结果

tutor_id	tutor_name	course_id	course_name
1	张三	1	语文
2	李四	2	数学

专家讲解

```
//MyBatis 中 collection 和 association 的区别？
public class A{
    private B b1;
    private List<B> b2;
}
在映射 b1 属性时用 association 标签,（一对一的关系）
映射 b2 时用 collection 标签（一对多的关系）
```

小结

本章主要讲的 resultMap 元素是 MyBatis 中最重要、最强大的元素。它可以让你远离 90% 的需要从结果集中取出数据的 JDBC 代码的那个东西，而且在某些情形下允许做一些

JDBC 不支持的事。

但是在现实的项目中进行数据库建模时，我们要遵循数据库设计范式的要求，会对现实中的业务模型进行拆分，封装在不同的数据表中，表与表之间存在着一对多或是多对多的对应关系。进而，我们对数据库的增删改查操作的主体，也就从单表变成了多表，因此掌握 MyBatis 处理表之间的关系尤为重要。

经典面试题

（1）MyBatis resultMap 用在什么情况下？
（2）为什么 MyBatis 的 xml 文件中可以有多个 resultMap?
（3）MyBatis 配置文件 resultMap 映射什么时候可写可不写？
（4）MyBatis 中关于 resultType 和 resultMap 的区别是什么？
（5）MyBatis 的 resultMap 能映射多个类吗？

跟我上机

在数据库中创建三张数据表，分别是登录账号表、客户表和销售人员表：
①实现客户与登录用户一对一关系；
②实现销售与登录用户一对一关系；
③实现销售与客户多对多关系。

第 5 章　关联关系和动态查询

本章要点（学会后请在方框中打钩）：

- □ MyBatis 一对一查询
- □ MyBatis 一对多查询
- □ MyBatis 动态标签 -<if><choose><when><otherwise><where><trim><set> 和 <foreach> 的用法

5.1 MyBatis 一对一查询

5.1.1 实例演示：用户登录表和用户信息表的关系

1）编写 User.java

User.java 的具体代码如下。

```
import java.io.Serializable;
@SuppressWarnings("serial")
public class User implements Serializable{
        private String id;
        private String password;
        private UserInfo userInfo;
        //setter/getter 略
}
```

2）创建 UserInfo.java

UserInfo.java 的具体代码如下。

```
@SuppressWarnings("serial")
public class UserInfo implements Serializable {
       private String userid;
        private String department;
        private String position;
        private String mobile;
        private String gender;
        private String email;
}
```

3）创建 UserInfoMapper.xml

UserInfoMapper.xml 的具体代码如下。

```
<?xml version="1.0" encoding="UTF-8" ?>
<!DOCTYPE mapper
PUBLIC "-//mybatis.org//DTD Mapper 3.0//EN"
"http://mybatis.org/dtd/mybatis-3-mapper.dtd">
<mapper namespace="com.iss.mapper.userInfoMapper">
      <resultMap type="userInfo" id="UserInfoResult">
```

```
            <id property="userid" column="userid" />
            <result property="department" column="department" />
            <result property="position" column="position" />
            <result property="mobile" column="mobile" />
            <result property="gender" column="gender" />
            <result property="email" column="email" />
        </resultMap>
        <select id="findUserInfoById" parameterType="String" resultMap="UserInfoResult">
            select * from userinfo where userid=#{id}
        </select>
</mapper>
```

4）创建 UserMapper.xml 中的 resultMap 配置

创建 UserMapper.xml 中的 resultMap 配置，具体代码内容如下。

```
<resultMap type="user" id="UserResult">
            <id property="id" column="id" />
            <result property="password" column="password" />
             <association property="userInfo" column="userid"select=" com.iss.mapper.userInfoMapper.findUserInfoById"></association>
        </resultMap>
```

5）将 UserInfoMapper.xml 加入 mybatis-config.xml

将 UserInfoMapper.xml 加入 mybatis-config.xml，具体代码内容如下。

```
<mappers>
        <mapper resource="mappers/UserMapper.xml" />
        <mapper resource="mappers/UserInfoMapper.xml" />
</mappers>
```

5.2 MyBatis 一对多查询

上一节演示了一个人对应一条个人信息，但是如果要根据部门来查询用户，就是典型的一对多关系。为了演示一对多关系查询的实现，我们需要现在数据库中创建一个部门表。

1）新增 Departments.java 文件

新增 Departments.java 文件的具体代码内容如下。

```
@SuppressWarnings("serial")
public class Departments implements Serializable{
        private String id;
        private String departmentName;
        private List<UserInfo> userInfos;
       //setter/getter 略
}
```

2）新增 DepartmentsDaoMapper.xml 文件

新增 DepartmentsDaoMapper.xml 文件，具体代码内容如下。

```
<?xml version="1.0" encoding="UTF-8" ?>
<!DOCTYPE mapper
PUBLIC "-//mybatis.org//DTD Mapper 3.0//EN"
"http://mybatis.org/dtd/mybatis-3-mapper.dtd">
<mapper namespace="com.iss.mapper.departmentsDaoMapper">
        <resultMap type="Departments" id="DepartmentsResult">
                <id property="id" column="id" />
                <result property="departmentName" column="departmentName" />
                <collection property="userInfos" column="id" select="com.iss.mapper. userInfoMapper.findUserInfoByDepartmentId"></collection>
        </resultMap>
        <select id="findDepartmentById" parameterType="String" resultMap="DepartmentsResult">
                select * from departments where id=#{id}
        </select>
</mapper>
```

3）修改 UserInfoMapper.xml 文件

在 UserInfoMapper.xml 文件中新增如下内容。

```
<select id="findUserInfoByDepartmentId" parameterType="String" resultMap="UserInfoResult">
                select * from userinfo where department=#{id}
        </select>
```

我们设计的表结构关系是 user 表→ userinfo 表→ departments 表。换句话说，我们使用 user 表中的变量是无法查到部门表的信息的。另外，我们实际使用中也会遇到这样的例子，人员的属性中存在多个集合属性，如基本信息是一个集合，兴趣爱好是一个集合，等等。有时为了完整描述一个对象，需要使用多个集合属性进行描述。因此，这里将演示一个对象中多个集合的查询操作的实现。

我们要在 User 中增加 department 属性，注意，我们只修改对象，并没有修改数据库。

```
@SuppressWarnings("serial")
public class User implements Serializable{

    private String id;
    private String password;
    private UserInfo userInfo;
    private Departments department;
//....set
//...get
//... 构造函数
//..toString( )
}
```

4）修改 UserMapper.xml 中的 resultMap

修改 UserMapper.xml 中的 resultMap，修改之后的代码内容如下。

```
<resultMap type="user" id="UserResult">
            <id property="id" column="id" />
            <result property="password" column="password" />
            <association property="userInfo" column="userid"
                  select="com.iss.mapper.userMapper.findUserInfoById"></association>
            <association property="department" column="department"
                  select="com.iss.mapper. departmentsDaoMapper.findDepartmentById"></association>
      </resultMap>
```

5.3 MyBatis 动态查询：<if>

修改 UserInfoMapper.xml 文件，代码内容如下。

```
<?xml version="1.0" encoding="UTF-8" ?>
<!DOCTYPE mapper
PUBLIC "-//mybatis.org//DTD Mapper 3.0//EN"
"http://mybatis.org/dtd/mybatis-3-mapper.dtd">
<mapper namespace="com.iss.mapper.userInfoMapper">
      <resultMap type="userInfo" id="UserInfoResult">
```

```
        <id property="userid" column="userid" />
        <result property="department" column="department" />
        <result property="position" column="position" />
        <result property="mobile" column="mobile" />
        <result property="gender" column="gender" />
        <result property="email" column="email" />
    </resultMap>
    <select id="findUserInfoByParams" parameterType="Map" resultMap="UserInfoResult">
        select * from userinfo
        where department=#{department}
        <if test="gender!=null">
            and gender = #{gender}
        </if>
        <if test="position!=null">
            and position like #{position}
        </if>
    </select>
</mapper>
```

专家讲解

这里的 where 之后至少有一个 department 条件。而 gender 和 position 条件会动态判断。这里的处理逻辑如下：

（1）假设没有传入动态参数，那么只会按照 department 条件查询；

（2）如果传入了动态的参数，则就在 department 条件上加入动态参数查询，具体代码内容如下。

```
@Test
public void testSelet( ) {
    Map<String,Object> map=new HashMap<String,Object>( );
    map.put("department", "2");
    map.put("gender", "1");
    map.put("position", "%售%");
    UserInfoDao userInfo = sqlSession.getMapper(UserInfoDao.class);
    List<UserInfo> UIList= userInfo.findUserInfoByParams(map);
    for(UserInfo ui:UIList){
        System.out.println(ui.toString( ));
    }
}
```

5.4 MyBatis 动态查询：<choose><when><otherwise>

在 UserInfoMapper.xml 中增加如下代码内容。

```
<select id="findUserInfoByOneParam" parameterType="Map" resultMap="UserInfoResult">
        select * from userinfo
        <choose>
            <when test="searchBy=='department'">
                where department=#{department}
            </when>
            <when test="searchBy=='position'">
                where position=#{position}
            </when>
            <otherwise>
                where gender=#{gender}
            </otherwise>
        </choose>
    </select>
```

专家讲解

(1)请注意这里的 select*fromuserinfo 之后没有再写 where 语句。

(2)<choose><when><otherwise> 配合使用是等价于 java 中的以下语句形式。

```
if(...){
....
}else if(...){
...
}else{
....
}
```

(3)请注意，每一个等号后面的参数都带有单引号，这是这种用法必须有的。否则，直接抛出异常。

(4)针对 department，这里有另外一种情况，那就是判断 null 的键值对，具体代码内容如下。

```
<select id="findUserInfoByOneParam" parameterType="Map" resultMap="UserInfoResult">
        select * from userinfo
        <choose>
            <when test="department!=null">
                where department=#{department}
```

```
                </when>
                <when test="position!=null">
                    where position=#{position}
                </when>
            </choose>
        </select>
```

5.5 MyBatis 动态查询：<where><trim><set>

前面在讲述 <if> 时说到至少有一个 where 固定条件，并且在缺失固定条件时 SQL 的执行结果是有 0 个符合条件的数据。可现实情况是，我们有时候无法确定输入条件中是不是至少有一个有效。这个问题我们可以用标签解决。

5.5.1 <where> 标签

修改 UserInfoMapper.xml，具体代码内容如下。

```
<select id="findUserInfoByUnoQuantity" parameterType="Map"
            resultMap="UserInfoResult">
            select * from userinfo
            <where>
                <if test="department!=null">
                department like #{department}
                </if>
                <if test="gender!=null">
                    AND gender=#{gender}
                </if>
                <if test="position!=null">
                    AND position like #{position}
                </if>
            </where>
    </select>
```

专家讲解

（1）select 之后没有直接写 SQL 语句的 where，而是使用 <where> 标签。

（2）按照标准写法，第一个 <if> 标签内的 AND 应该不写，但是，就算开发中书写也不会报错。这就是 where 标签帮助我们自动移除了第一个 AND 链接。但是，第二个之后的 <if> 标签内，必须有 AND 链接。

（3）如果没有一个条件符合，则返回所有条目。

（4）结论：where 元素知道只有在一个以上的 <if> 条件有值的情况下才去插入“WHERE”子句。而且，若内容是“AND”或“OR”开头的，where 元素也知道如何将他们去除。

5.5.2　<trim> 标签

<trim> 标签的功能与 <where> 类似，并且额外提供了前缀后缀功能，现可修改 Mapper 文件具体代码内容如下。

```
<select id="findUserInfoByTrim" parameterType="Map"
            resultMap="UserInfoResult">
            select * from userinfo
            <trim prefix="where" prefixOverrides="and|or">
                <if test="department!=null">
                    AND department like #{department}
                </if>
                <if test="gender!=null">
                    AND gender=#{gender}
                </if>
                <if test="position!=null">
                    AND position like #{position}
                </if>
            </trim>
</select>
```

专家讲解

（1）我们使用 <trim> 替代 <where> 标签。

（2）属性“prefix”表示：加入前缀 where。

（3）属性“prefixOverrides”表示：自动覆盖第一个“and”或者“or”。

5.5.3　<set> 标签

<set> 标签用于 update 语句，现修改 Mapper 文件，具体代码内容如下。

```
<update id="updateUserInfoBySet" parameterType="userInfo">
            update userInfo
            <set>
                <if test="mobile!=null">
```

```
                mobile=#{mobile},
            </if>
            <if test="gender!=null">
                gender=#{gender},
            </if>
            <if test="position!=null">
                position = #{position},
            </if>
        </set>
        where userid=#{userid}
</update>
```

专家讲解

（1）SQL 语句的 set 被 <set> 标签替代。

（2）每个 <if> 中语句最后都带有逗号，写过 SQL 语句的同学就一定知道，最后的逗号是不能有的，因此这里的 <set> 标签能够自动移除最后一个 <if> 中的逗号。

（3）<trim> 是一个非常强大的标签，因此也可以通过 <trim> 来实现 <set> 的功能，示例代码如下：（这种写法的运行效果与 <set> 等价）

```
<update id="updateUserInfoBySet" parameterType="userInfo">
        update userInfo
        <trim prefix="SET" suffixOverrides=",">
            <if test="mobile!=null">
                mobile=#{mobile},
            </if>
            <if test="gender!=null">
                gender=#{gender},
            </if>
            <if test="position!=null">
                position = #{position},
            </if>
        </trim>
        where userid=#{userid}
</update>
```

5.6 MyBatis 动态查询：<foreach>

使用 MyBatis 提供的 <foreach> 标签可以实现某些循环增改删查的需求。

<foreach> 中可以供我们使用的属性如表 1 所示。

表 1　<foreach> 属性列表

属性	描述
item	循环体中的具体对象，支持属性的点路径访问，如 item.age,item.info.details。 具体说明：在 list 和数组中 item 是其中的对象，在 map 中是 value。 该参数为必选
collection	要做 foreach 的对象，作为入参时，List<?> 对象默认用 list 代替作为键，数组对象有 array 代替作为键，Map 对象没有默认的键。 当然在作为入参时可以使用 @Param("keyName") 来设置键，设置 keyName 后，list,array 将会失效。除了入参这种情况外，还有一种作为参数对象的某个字段的时候。举个例子： 如果 User 有属性 Listids。入参是 User 对象，那么这个 collection="ids" 如果 User 有属性 Idsids; 其中 Ids 是个对象，Ids 有个属性 Listid; 入参是 User 对象，那么 collection="ids.id" 上面只是举例，具体 collection 等于什么，就看你想对那个元素做循环。 该参数为必选
separator	元素之间的分隔符，例如在 in( ) 的时候，separator="," 会自动在元素中间用“,“隔开，避免手动输入逗号导致 sql 错误，如 in(1,2,) 这样。该参数可选
open	foreach 代码的开始符号，一般设置为“(“和 close=")" 合用。常用在 in( ),values( ) 时。该参数可选
close	foreach 代码的关闭符号，一般设置为“)“和 open="(" 合用。常用在 in( ),values( ) 时。该参数可选
index	在 list 和数组中 ,index 是元素的序号，在 map 中，index 是元素的 key，该参数可选

5.6.1　参数为数组 Array

UserTest 的单元测试方法代码内容如下。

```
@Test
    public void testForEachArray( ) {
        try {
            String[] sa = new String[]{"admin","customer","customer2"};
            UserInfoDao userInfo = sqlSession.getMapper(UserInfoDao.class);
            List<UserInfo> UIList = userInfo.selectUserInfoByForEachArray(sa);
            for (UserInfo ui : UIList) {
            System.out.println(ui.toString( ));
            }
        } catch (Exception e) {
```

```
                e.printStackTrace( );
            }
        }
```

UserInfoMapper.xml 的代码内容如下。

```
<?xml version="1.0" encoding="UTF-8" ?>
<!DOCTYPE mapper
PUBLIC "-//mybatis.org//DTD Mapper 3.0//EN"
"http://mybatis.org/dtd/mybatis-3-mapper.dtd">
<mapper namespace="com.iss.mapper.userInfoMapper">
    <resultMap type="userInfo" id="UserInfoResult">
        <id property="userid" column="userid" />
        <result property="department" column="department" />
        <result property="position" column="position" />
        <result property="mobile" column="mobile" />
        <result property="gender" column="gender" />
        <result property="email" column="email" />
    </resultMap>
<select id="selectUserInfoByForEachArray" resultMap="UserInfoResult">
        select * from userinfo
        <if test="array!=null">
            where userid in
            <foreach item="item" collection="array" index="index" open="("
                separator="," close=")">
                #{item}
            </foreach>
        </if>
    </select>
</mapper>
```

专家讲解

（1）这里的 if 标签能够判断出数组是否为空。但是没有判断出数组中的内容是否为空。即，当数组中有内容时或者数组为 null 时，该 SQL 语句能够正常执行，但是如果数组为空的话，if 判断的结果为真，但 foreach 执行 0 次。这种情况下，MyBatis 会组装出 1 条错误的 SQL 语句。换句话说这里 if 是多余的。

（2）这里 collection 属性配置为”array“

（3）index 在这条语句中未使用，所以是可以缺省的。移除也不会引起错误

（4）open，close 只会在开始与结尾出现一次。

（5）separator 会使用配置的“，”来每次间隔 <foreach> 标签内的内容。

（6）#{item} 中的 item 必须与 <foreach> 中的 item 属性的值 "item" 保持一致。

5.6.2 参数为 List

在 UserInfoDaoMapper.xml 中，增加如下内容。

```
<select id="selectUserInfoByForEachList" resultMap="UserInfoResult">
        select * from userinfo
        <if test="list!=null">
            where userid in
            <foreach item="item" collection="list" index="index" open="("
                separator="," close=")">
                #{item}
            </foreach>
        </if>
</select>
```

专家讲解

（1）与上面类似：这里的 if 标签能够判断出 List 是否为空。但是没有判断出 List 中的内容是否为空。即，当 List 中有内容时或者 List 为 null 时，该 SQL 语句能够正常执行，但是如果 List 为空的话，if 判断的结果为真，但 foreach 执行 0 次。这种情况下，MyBatis 会组装出 1 条错误的 SQL 语句。换句话说这里 if 是多余的。

（2）这里 collection 属性配置为“list”。

（3）index 在这条语句中未使用，所以是可以缺省的。移除也不会引起错误。

5.6.3 参数为 Map

在 UserInfoDaoMapper.xml 中，增加如下内容。

```
<select id="selectUserInfoByForEach" parameterType="Map"
        resultMap="UserInfoResult">
        select * from userinfo
        <if test="userids!=null">
            where userid in
            <foreach item="ParamsId" collection="userids" index="index"
                open="(" separator="," close=")">
                #{ParamsId}
```

```
                    </foreach>
               </if>
    </select>
```

专家讲解

（1）与上面类似：这里的 if 标签能够判断出 map 中否包含 userids 这个 key，其对应的 value 可以任意。但是没有判断出 value 中的内容是否为空。即，当 value 中有内容时或者 key 不存在时，该 SQL 语句能够正常执行，但是如果 value 为空的话，if 判断的结果为真，但 foreach 执行 0 次。这种情况下，MyBatis 会组装出 1 条错误的 SQL 语句。换句话说这里 if 是多余的。

（2）这里 collection 属性配置为 map 中 list 对应的 key 的值，而不是 value 或者"list"

（3）注意这里的 #{ParamsId} 是与属性 item 中的 ParamsId 对应的

5.7 MyBatis 动态查询：<sql>

SQL 片段的意思就是将 SQL 的部分语句单独定义出来。然后在使用的时候引入，如以下示例。

```
<mapper namespace="com.iss.model.IUserOperation">
<select id="findUserList" parameterType="com.iss.model.UserQueryVO"
  resultType="com.iss.model.UserCustom">
  select * from t_user
  <!-- where 可以自动去掉第一个条件的 and -->
  <where>
     <!-- sql 片段引入 -->
     <include refid="query_user_where"></include>
  </where>
</select>
<!-- sql 片段 -->
<sql id="query_user_where">
<if test="userCustom!=null">
  <if test="userCustom.sex!=null and userCustom.sex!=''">
and t_user.sex=#{userCustom.sex}
     </if>
  <if test="userCustom.username!=null and userCustom.username!=''and t_user.user-
name like '%${userCustom.username}%'
```

```
17.     </if>
18. </if>
19.  </sql>
20. </mapper>
```

小结

MyBatis 的一个强大特性是它的动态 SQL 能力。如果你有使用 JDBC 或其他相似框架的经验，你就明白条件串联 SQL 字符串在一起是多么地痛苦，确保不能忘了空格或者在列表的最后的省略逗号，动态 SQL 可以彻底处理这种痛苦。

通常使用动态 SQL 不可能是独立的一部分，MyBatis 当然使用一种强大的动态 SQL 语言来改进这种情形，这种语言可以被用在任意映射的 SQL 语句中。

动态 SQL 元素和使用 JSTL 或其他相似的基于 XML 的文本处理器相似，在 MyBatis 之前的版本中，有很多元素需要了解，MyBatis3 大大地提升了它们，现在用不到原先一半的元素就能工作了，MyBatis 采用功能强大的基于 OGNL 的表达式来消除其他元素。

MyBatis 的动态 SQL 功能正是为了解决这种问题，其通过 if, choose, when, otherwise, trim, where, set, foreach 标签，可组合成非常灵活的 SQL 语句，从而提高开发人员的效率。

纵观 MyBatis 的动态 SQL，强大而简单。

经典面试题

（1）MyBatis 的动态 SQL 有什么特点？

（2）MyBatis 写 SQL 能 if 中套 if 吗？

（3）MyBatis 的 if test 标签 java 怎么使用？

（4）MyBatis 里面的 foreach 怎么循环 list?

（5）MyBatis 中 foreach 如何获取当前循环次数？

（6）MyBatis 动态查询可以传集合参数吗？

（7）MyBatis 怎样实现 MySQL 动态分页？

跟我上机

使用 MyBatis 动态条件查询完成模糊搜索和分页查询。

要求：数据库表自行创建，并进行如下操作。

①尝试使用一对多的关系。

②尝试使用多对多的关系。

第 6 章　MyBatis 注解配置实现 CURD

本章要点（学会后请在方框中打钩）：

□ 掌握实现注解配置对单表进行 CRUD 操作

□ 掌握采用注解配置方式实现一对多关系的查询操作

□ 在 Java Web 项目中熟练注解使用

前面的例子讲解了如何使用 XML 配置方式操作 MyBatis 实现 CRUD，但是大量的 XML 配置文件的编写是非常烦琐的。因此 MyBatis 也提供了基于注解的配置方式。

6.1 了解 MyBatis 注解

MyBatis 可以利用 SQL 映射文件来配置，也可以利用 Annotation 来设置。MyBatis 提供的一些基本注解如表 1 所示。

表 1 基本注解

注解	目标	相应的 XML	描述
@CacheNamespace	类	<cache>	为给定的命名空间（比如类）配置缓存。属性：implemetation,eviction,flushInterval , size 和 readWrite
@CacheNamespaceRef	类	<cacheRef>	参照另外一个命名空间的缓存来使用。属性：value，也就是类的完全限定名
@ConstructorArgs	方法	<constructor>	收集一组结果传递给对象构造方法。属性：value，是形式参数的数组
@Arg	方法	<arg><idArg>	单独的构造方法参数，是 ConstructorArgs 集合的一部分。属性：id，column，javaType，typeHandler。id 属性是布尔值，来标识用于比较的属性，和 <idArg>XML 元素相似
@TypeDiscriminator	方法	<discriminator>	一组实例值被用来决定结果映射的表现。属性：Column，javaType，jdbcTypetypeHandler，cases。cases 属性就是实例的数组
@Case	方法	<case>	单独实例的值和它对应的映射。属性：value，type，results 。Results 属性是结果数组，因此这个注解和实际的 ResultMap 很相似，由下面的 Results 注解指定
@Results	方法	<resultMap>	结果映射的列表，包含了一个特别结果列如何被映射到属性或字段的详情。属性：value ，是 Result 注解的数组
@Result	方法	<result> <id>	在列和属性或字段之间的单独结果映射。属性：id，column，property，javaType，jdbcType，type Handler，one，many。id 属性是一个布尔值，表示了应该被用于比较的属性。one 属性是单独的联系，和 <association> 相似，而 many 属性是对集合而言的，和 <collection> 相似

续表

注解	目标	相应的 XML	描述
@One	方法	<association>	复杂类型的单独属性值映射。属性：select，已映射语句（也就是映射器方法）的完全限定名，它可以加载合适类型的实例。注意：联合映射在注解 API 中是不支持的
@Many	方法	<collection>	复杂类型的集合属性映射。属性：select，是映射器方法的完全限定名，它可加载合适类型的一组实例。注意：联合映射在 Java 注解中是不支持的
@Options	方法	映射语句的属性	这个注解提供访问交换和配置选项的宽广范围，它们通常在映射语句上作为属性出现。而不是将每条语句注解变复杂，Options 注解提供连贯清晰的方式来访问它们。属性：useCache=true， flushCache=false， resultSetType=FORWARD_ONLY， statementType=PREPARED， fetchSize= -1，timeout=-1 ， useGeneratedKeys=false ， keyProperty="id"。 理解 Java 注解是很重要的，因为没有办法来指定"null"作为值。因此，一旦你使用了 Options 注解，语句就受所有默认值的支配。要注意什么样的默认值来避免不期望的行为
@Insert @Update @Delete	方法	<insert> <update> <delete>	这些注解中的每一个代表了执行的真实 SQL。它们每一个都使用字符串数组（或单独的字符串）。如果传递的是字符串数组，它们由每个分隔它们的单独空间串联起来。属性：value，这是字符串数组用来组成单独的 SQL 语句
@InsertProvider @UpdateProvider @DeleteProvider @SelectProvider	方法	<insert> <update> <delete> <select> 允许创建动态 SQL	这些可选的 SQL 注解允许你指定一个类名和一个方法在执行时来返回运行的 SQL。基于执行的映射语句，MyBatis 会实例化这个类，然后执行由 provider 指定的方法。这个方法可以选择性的接受参数对象作为它的唯一参数，但是必须只指定该参数或者没有参数。属性：type，method。type 属性是类的完全限定名。method 是该类中的那个方法名
@Param	参数	N/A	当映射器方法需多个参数，这个注解可以被应用于映射器方法参数来给每个参数一个名字。否则，多参数将会以它们的顺序位置来被命名。比如 #{1}，#{2} 等，这是默认的。使用 @Param("person")，SQL 中参数应该被命名为 #{person}

这些注解都是运用到传统意义上映射器接口中的方法、类或者方法参数中的。

6.2 综合实例演示

下面来演示一下使用接口加注解来实现 CRUD 的的例子。演示操作数据库表如图 1 所示。

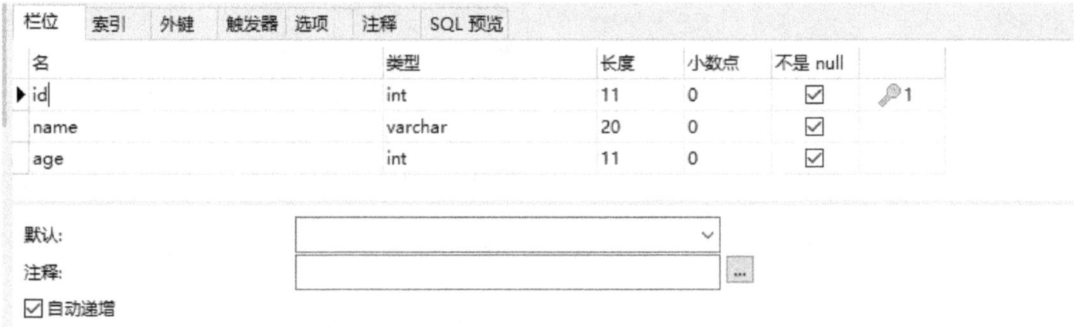

栏位 索引 外键 触发器 选项 注释 SQL 预览

名	类型	长度	小数点	不是 null	
id	int	11	0	☑	1
name	varchar	20	0	☑	
age	int	11	0	☑	

默认:

注释:

☑自动递增

图 1 演示操作数据库 Users 表

6.2.1 定义 SQL 映射的接口

UserMapperI 接口的代码内容如下：

```
package com.iss.mapper;
public interface UserMapperI {
        // 使用 @Insert 注解指明 add 方法要执行的 SQL
        @Insert("insert into users(name, age) values(#{name}, #{age})")
        public int add(User user);

        // 使用 @Delete 注解指明 deleteById 方法要执行的 SQL
        @Delete("delete from users where id=#{id}")
        public int deleteById(int id);

        // 使用 @Update 注解指明 update 方法要执行的 SQL
        @Update("update users set name=#{name},age=#{age} where id=#{id}")
        public int update(User user);

        // 使用 @Select 注解指明 getById 方法要执行的 SQL
        @Select("select * from users where id=#{id}")
        public User getById(int id);
```

```
    // 使用 @Select 注解指明 getAll 方法要执行的 SQL
    @Select("select * from users")
    public List<User> getAll( );
}
```

需要说明的是，我们不需要针对 UserMapperI 接口去编写具体的实现类代码，这个具体的实现类由 MyBatis 帮我们动态构建出来，我们只需要直接拿来使用即可。

6.2.2 在 myBatisConfig.xml 文件中注册映射接口

```
<?xml version="1.0" encoding="UTF-8"?>
<!DOCTYPE configuration PUBLIC "-//mybatis.org//DTD Config 3.0//EN"
"http://mybatis.org/dtd/mybatis-3-config.dtd">
<configuration>
    <properties resource="jdbc.properties"></properties>
    <environments default="development">
        <environment id="development">
            <transactionManager type="JDBC"></transactionManager>
            <dataSource type="POOLED">
                <property name="driver" value="${driver}" />
                <property name="url" value="${url}" />
                <property name="username" value="${username}" />
                <property name="password" value="${password}" />
            </dataSource>
        </environment>
    </environments>
        <mappers>
            <mapper class="com.iss.mapper.ITeacherMapper" />
        </mappers>
</configuration>
```

jdbc.properties 配置文件内容如下：

```
driver=org.gjt.mm.mysql.Driver
url=jdbc:mysql://localhost:3306/mybatisdb?useUnicode=true&characterEncoding=utf-8
username=root
password=root
```

6.2.3 编写 MyBatisUtil 工具类

```
public class MyBatisUtil {
    /**
     * 获取 SqlSessionFactory
     * @return SqlSessionFactory
     */
    public static SqlSessionFactory getSqlSessionFactory( ) {
        String resource = "../../../myBatisConfig.xml";
        InputStream is = MyBatisUtil.class.getClassLoader( ).getResourceAsStream(resource);
        SqlSessionFactory factory = new SqlSessionFactoryBuilder( ).build(is);
        return factory;
    }

    /**
     * 获取 SqlSession
     * @return SqlSession
     */
    public static SqlSession getSqlSession( ) {
        return getSqlSessionFactory( ).openSession( );
    }

    /**
     * 获取 SqlSession
     * @param isAutoCommit
     * true 表示创建的 SqlSession 对象在执行完 SQL 之后会自动提交事务 false
     * 表示创建的 SqlSession 对象在执行完 SQL 之后不会自动提交事务，这时就需要我们手动调用 sqlSession.commit( ) 提交事务
     * @return SqlSession
     */
    public static SqlSession getSqlSession(boolean isAutoCommit) {
        return getSqlSessionFactory( ).openSession(isAutoCommit);
    }
}
```

6.2.4 编写单元测试类

1）测试添加功能

```
@Test
public void testAdd( ) {
        SqlSession sqlSession = MyBatisUtil.getSqlSession(true);
        // 得到 UserMapperI 接口的实现类对象，
        // UserMapperI 接口的实现类对象由 sqlSession.getMapper(UserMapperI.class) 动态构建出来
        UserMapperI mapper = sqlSession.getMapper(UserMapperI.class);
        User user = new User( );
        user.setName(" 张建军 ");
        user.setAge(40);
        int add = mapper.add(user);
        // 使用 SqlSession 执行完 SQL 之后需要关闭 SqlSession
        sqlSession.close( );
        System.out.println(add);
}
```

2）测试更新功能

```
@Test
public void testUpdate( ) {
        SqlSession sqlSession = MyBatisUtil.getSqlSession(true);
        // 得到 UserMapperI 接口的实现类对象，UserMapperI 接口的实现类对象由 sqlSession.getMapper(UserMapperI.class) 动态构建出来
        UserMapperI mapper = sqlSession.getMapper(UserMapperI.class);
        User user = new User( );
        user.setId(1);
        user.setName(" 何晶 ");
        user.setAge(35);
        // 执行修改操作
        int retResult = mapper.update(user);
        // 使用 SqlSession 执行完 SQL 之后需要关闭 SqlSession
        sqlSession.close( );
        System.out.println(retResult);
}
```

3）测试删除功能

```
@Test
    public void testDelete( ) {
        SqlSession sqlSession = MyBatisUtil.getSqlSession(true);
      // 得到 UserMapperI 接口的实现类对象，UserMapperI 接口的实现类对象由 sqlSession.getMapper(UserMapperI.class) 动态构建出来
      UserMapperI mapper = sqlSession.getMapper(UserMapperI.class);
      // 执行删除操作
      int retResult = mapper.deleteById(1);
      // 使用 SqlSession 执行完 SQL 之后需要关闭 SqlSession
      sqlSession.close( );
      System.out.println(retResult);
    }
```

4）测试 ById 查询功能

```
@Test
    public void testGetUser( ) {
        SqlSession sqlSession = MyBatisUtil.getSqlSession( );
        // 得到 UserMapperI 接口的实现类对象，UserMapperI 接口的实现类对象由 sqlSession.getMapper(UserMapperI.class) 动态构建出来
        UserMapperI mapper = sqlSession.getMapper(UserMapperI.class);
        // 执行查询操作，将查询结果自动封装成 User 返回
        User user = mapper.getById(2);
        // 使用 SqlSession 执行完 SQL 之后需要关闭 SqlSession
        sqlSession.close( );
        System.out.println(user);
    }
```

5）测试查询全部数据

```
@Test
    public void testGetAll( ) {
        SqlSession sqlSession = MyBatisUtil.getSqlSession( );
        // 得到 UserMapperI 接口的实现类对象，UserMapperI 接口的实现类对象由 sqlSession.getMapper(UserMapperI.class) 动态构建出来
        UserMapperI mapper = sqlSession.getMapper(UserMapperI.class);
        // 执行查询操作，将查询结果自动封装成 List<User> 返回
        List<User> lstUsers = mapper.getAll( );
```

```
            // 使用 SqlSession 执行完 SQL 之后需要关闭 SqlSession
            sqlSession.close( );
            System.out.println(lstUsers);
    }
```

以上的相关代码已全部测试通过（亲测），结果图省略。

6.3 结果映射：@ResultMap

在 xml 配置文件中，将查询结果和 JavaBean 属性映射起来的标签是 <resultMap>。对应的是 @Results 注解，可以解决数据库表字段和 POJO 类的属性名称不一致的情况

```
@Select("select * from user")
@Results({ @Result(id = true, column = "id", property = "id"),
            @Result(column = "username", property = "user_name"),
            @Result(column = "city", property = "city") })
public List<User> selectAll( ) throws Exception;
```

@Results 注解没办法复用。譬如 public User selectById(int id) throws Exception 也要用到同样的 @Results 注解，但还是要重新写一个一模一样的 @Results

```
@Select("select * from user where id=#{id}")
@Results({ @Result(id = true, column = "id", property = "id"),
            @Result(column = "username", property = "user_name"),
            @Result(column = "city", property = "city") })
public User selectById(int id) throws Exception;
```

如果想使用可以复用的映射器，就选择 @ResultMap 注解。该注解依赖一个 xml 配置文件。 在接口文件同目录下新建一个 userMapper.xml 文件，并定义一个名为 userMap 的 resultMap。

```
<mapper namespace="com.iss.mapper.UserMapper">
    <!-- 自定义返回结果集 -->
    <resultMap id="userMap" type="com.iss.pojo.User">
        <id column="id" property="id" jdbcType="INTEGER" />
        <result property="user_name" column="username"></result>
        <result property="city" column="city"></result>
    </resultMap>
</mapper>
```

在 UserMapper.Java 中，使用 @ResultMap 引用名为 userMap 的 resultMap，实现复用。

```
@Select("select * from user where id=#{id}")
@ResultMap("com.iss.mapper.UserMapper.userMap")
public User selectById(int id) throws Exception;
@Select("select * from user")
@ResultMap("com.iss.mapper.UserMapper.userMap")
public List<User> selectAll( ) throws Exception;
```

6.4 综合实例演示：注解实现表的关联关系

项目结构图如图 2 所示。

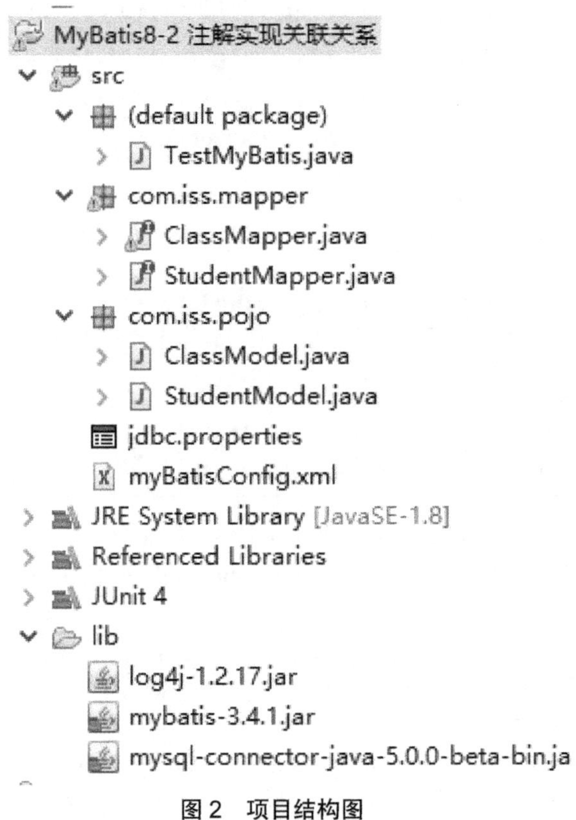

图 2 项目结构图

6.4.1 一对一关系

MyBatis 提供了 @One 注解来配合 @Result 注解，从而实现一对一关联查询数据的加载。

如学生和班级的关系，一个学生只属于一个班级。

1）StudentModel.java

```
public class StudentModel {
        String sid;
        String name;
        String cid;
        ClassModel clazz;
//setter/getter 省略
}
```

2）StudentMapper.java

```
public interface StudentMapper {
        @Select("select * from tb_student where sid=#{sid}")
        @Results({
                @Result(column = "name", property = "name"),
                @Result(id = true, column = "sid", property = "sid"),
                @Result(column = "cid", property = "clazz", one =
 @One(fetchType=FetchType.LAZY,select =
 "com.iss.mapper.ClassMapper.findClassByCid")) })
        StudentModel findBySid(String sid);

        @Select("select * from tb_student where cid=#{cid}")
         List<StudentModel> findAllStuByCid(String cid);
}
```

6.4.2 一对多关系

MyBatis 提供了 @Many 注解来配合 @Result 注解，从而实现一对多关联查询数据的加载。

1）ClassModel.java

```
public class ClassModel {
        String cid;
        String cname;
        List<StudentModel> students;
//setter/getter 省略
}
```

2）ClassMapper.java

```
public interface ClassMapper {
        @Select("select * from tb_classinfo where cid=#{cid}")
        @Results({
                @Result(id = true, column = "cid", property = "cid"),
                @Result(column = "cname", property = "cname"),
                @Result(column = "cid", property = "students", many = @Many(select = "com.iss.mapper.StudentMapper.findAllStuByCid")) })
        ClassModel findClassByCid(String cid);
}
```

6.4.3 编写测试方法

```
@Test
        public void test( ) {
                SqlSessionFactoryBuilder sfb = new SqlSessionFactoryBuilder( );
                SqlSessionFactory ss = sfb.build(this.getClass( ).getResourceAsStream("myBatisConfig.xml"));
                SqlSession session = ss.openSession( );
                StudentMapper stuMapper = session.getMapper(StudentMapper.class);
                StudentModel stuModel = stuMapper.findBySid("2014001");
                System.out.println(stuModel.getSid( ) + "," + stuModel.getName( ) + "," + stuModel.getClazz( ).getCname( ));

                ClassMapper sm = session.getMapper(ClassMapper.class);
                ClassModel stu = sm.findClassByCid("C2");
                System.out.print(stu.getCid( ) + "," + stu.getCname( ));
                List<StudentModel> list = stu.getStudents( );
                for (StudentModel studentModel : list) {
                        System.out.print(studentModel.getName( ) + ",");
                }
        }
```

练习：增加老师的实体类，考虑都是什么关系，并实现。

小结

MyBatis 支持使用注解来配置映射的 sql 语句，这样可以省掉映射器 xml 文件，简洁方便。但是比较复杂的 SQL 和动态 SQL 还是建议书写类配置文件。

经典面试题

（1）使用 MyBatis 注解有什么优点？
（2）Spring 整合 MyBatis 怎样配置注解？
（3）MyBatis 如何注解 resultMap 一对多关系？
（4）MyBatis 基于注解式的事务管理怎么配置？
（5）MyBatis 怎么实现对象参数和注解参数的同时传入？
（6）如何实现基于注解 MyBatis 动态拼写 SQL 语句？
（7）MyBatis 注解怎么解决字段名与属性名不同的问题？
（8）MyBatis 注解配置文件中统计总记录数 SQL 语句怎么写？

跟我上机

1. 在 Oracle/MySql 数据库下创建一个航班信息表 fightinfo 和 城市信息表 cityinfo，使用注解配置方式，完成图 3 所示功能。

flightinfo 表结构如表 2 所示。

表 2　flightinfo 表结构

字段名	类型	约束	描述
flightid	NUMBER(4)	主键	航班 id 序列增长
flightnum	VARCHAR2(10)	非空	航班号
flydate	DATE	非空	飞行日期
starttime	VARCHAR2(10)	非空	发出时间（例：8：00）
flytime	VARCHAR2(10)	非空	飞行时间（例：2 天）
startcity	NUMBER(4)	外键	始发地
endcity	NUMBER(4)	外键	目的地
seatnum	NUMBER(4)	非空	座位总数

cityinfo 表结构如表 3 所示。

表 3　cityinfo 表结构

字段名	类型	约束	描述
cityid	NUMBER(2)	主键	城市 id
cityname	VARCHAR2(10)	非空	城市名称

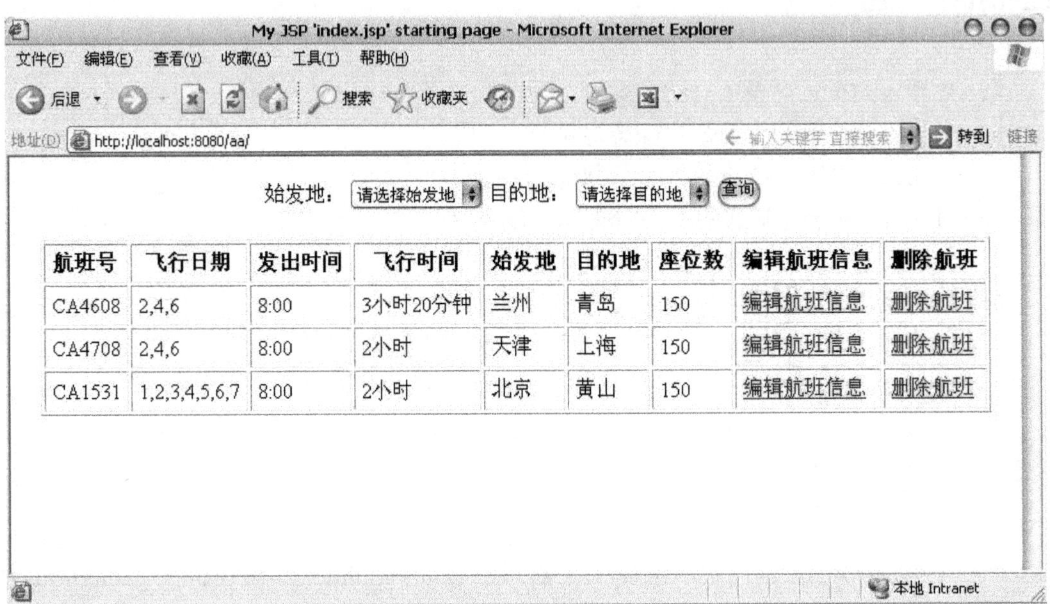

图 3　运行结果

第 7 章　MyBatis 分页查询

本章要点（学会后请在方框中打钩）：

- □ 了解 MySql 分页语句的写法
- □ 掌握分页通用工具类的创建
- □ 掌握逻辑分页及其实现
- □ 熟练掌握物理分页及其实现
- □ 结合 BootStrap 技术实现分页功能的编写

目前为止介绍的 MyBatis 的种种查询都是一次性查询出所有结果并返回给上层。但是在实际开发过程中大量数据存在的情况下，必须使用分页。所以现在将从逻辑分页、物理分页两种情况出发分别介绍这两种方式。

（1）首先，准备数据 (我们也可以直接沿用以前的数据库)，具体内容如图 1 所示。

Tch_ID	Tch_Name	Tch_Subj	Tch_Dept	Tch_Position	Tch_Class	Tch_Inspect
1	张三	语文	高一年级组	年级组长	1	是
3	王五	物理	高一年级组	班主任	1	否
4	赵六	化学	高一年级组	教师	1	否
5	孙七	英语	高一年级组	教师	1	否
6	林一	生物	高一年级组	班主任	1	否
7	林二	语文	高一年级组	教师	2	否
8	林三	历史	高二年级组	年级组长	2	否
9	林四	语文	高二年级组	年级副组长	2	否
10	林五	数学	高二年级组	班主任	3	否
11	林六	英语	高二年级组	教师	3	否
12	林七	语文	高二年级组	班主任	3	否

图 1　数据库 teacher 表

（2）创建本例中使用的工程，MyBatis9-2 物理分页，工程结构如图 2 所示。

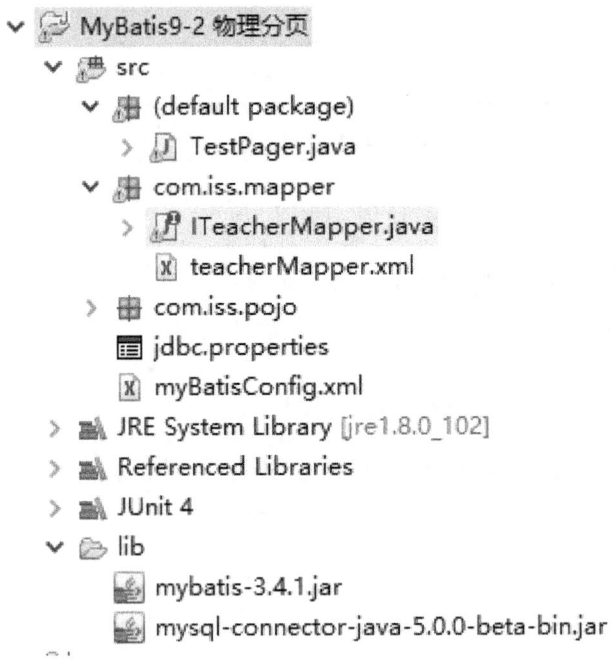

```
MyBatis9-2 物理分页
  src
    (default package)
      TestPager.java
    com.iss.mapper
      ITeacherMapper.java
      teacherMapper.xml
    com.iss.pojo
    jdbc.properties
    myBatisConfig.xml
  JRE System Library [jre1.8.0_102]
  Referenced Libraries
  JUnit 4
  lib
    mybatis-3.4.1.jar
    mysql-connector-java-5.0.0-beta-bin.jar
```

图 2　工程目录结构

（3）Teacher 为 POJO 对象，属性与数据库表对应即可。

```
public class Teacher {
int tch_id;
String tch_name;
String tch_subj;
String tch_dept;
String tch_position;
int tch_class;
String tch_inspect;
//setter 和 getter 省略
}
```

7.1 逻辑分页

逻辑分页虽然看起来实现了分页的功能，但实际上是将查询的所有结果放置在内存中，每次都从内存中获取。这种情况适用于数据量较少的情况。因此在实际开发中，基本不会使用到逻辑分页功能。

7.1.1 实例演示：XML 配置方式

（1）在 teacherMapper.xml 文件中增加对应查询语句，具体代码内容如下：

```
<?xml version="1.0" encoding="UTF-8"?>
<!DOCTYPE mapper PUBLIC "-//mybatis.org//DTD Mapper 3.0//EN" "http://mybatis.org/dtd/mybatis-3-mapper.dtd">
<mapper namespace="com.iss.mapper.teacherMapper">
<select id="selectTeacher" resultType="com.iss.pojo.Teacher">
    select * from teacher
</select>
</mapper>
```

（2）增加对应单元测试方法，具体代码内容如下：

```
@Test
public void testSqlectTeacher( ) {
    InputStream stream = this.getClass( ).getResourceAsStream("myBatisConfig.xml");
    SqlSessionFactory factory = new SqlSessionFactoryBuilder( ).build(stream);
```

```
        SqlSession sqlSession = factory.openSession(true);
        String statement = "com.iss.mapper.teacherMapper.selectTeacher";
        int start = 0;// 正确的叫法应该是 offset
        int pagesize = 5;// 正确的叫法应该是 limit
        RowBounds rowBounds = new RowBounds(start*pagesize, pagesize);
        List<Teacher> teacherList = sqlSession.selectList(statement, Teacher.class, rowBounds);
         System.out.println(" 第 " + (start + 1) + " 页 ");
        for (Teacher teacher : teacherList) {
            System.out.println(teacher.getTch_id( ) + "," + teacher.getTch_name( ) + "," + teacher.getTch_position( ));
        }
}
```

（3）运行单元测试方法，观察控制台输出，如图 3 所示。

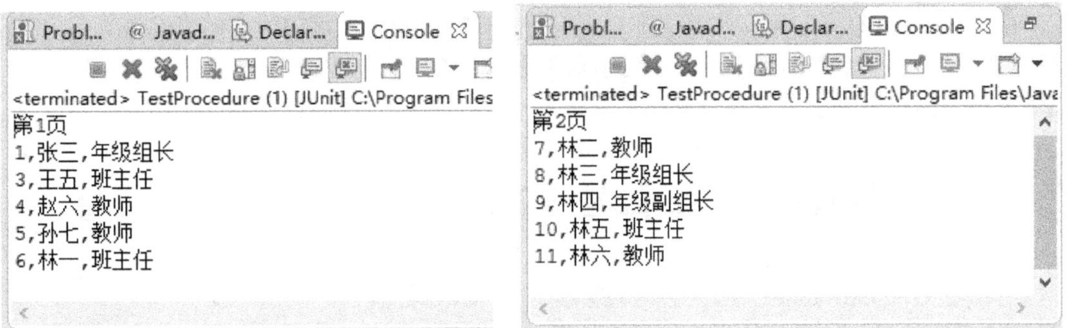

图 3　控制台输出

专家提醒

（1）List<Teacher> teacherList = sqlSession.selectList(statement, Teacher.class, rowBounds);// 这句中的 Teacher.class 必须要加上。

（2）注解配置方式是不能使用 RowBounds 类进行测试的。

（3）上面的这种方式在实际性能表现上存在隐患，各位读者最好还是不要使用，作为参考即可。

7.2　物理分页

这种分页方法从底层上就是每次只查询对应条目数量的数据，从而实现了真正意义上的分页。

7.2.1 实例演示:XML 配置方式

（1）在 teacherMapper.xml 文件中增加对应查询语句，具体代码内容如下：

```
<?xml version="1.0" encoding="UTF-8"?>
<!DOCTYPE mapper PUBLIC "-//mybatis.org//DTD Mapper 3.0//EN"
 "http://mybatis.org/dtd/mybatis-3-mapper.dtd">
<mapper namespace="com.iss.mapper.teacherMapper">
<select id="selectTeacher" parameterType="Map" resultType="com.iss.pojo.Teacher">
    select * from teacher
    <if test="offset!=null and pagesize!=null">
        limit #{offset},#{pagesize}
    </if>
</select>
</mapper>
```

（2）增加对应单元测试方法，具体代码内容如下：

```
@Test
public void testSqlectTeacher1( ) {
    InputStream stream = this.getClass( ).getResourceAsStream("myBatisConfig.xml");
    SqlSessionFactory factory = new SqlSessionFactoryBuilder( ).build(stream);
    SqlSession sqlSession = factory.openSession(true);
    String statement = "com.iss.mapper.teacherMapper.selectTeacher";
    int start = 0;// 正确的叫法应该是 offset
    int pagesize = 5;// 正确的叫法应该是 limit
    Map map = new HashMap( );
    map.put("offset", start * pagesize);
    map.put("pagesize", pagesize);
    List<Teacher> teacherList = sqlSession.selectList(statement, map);
    System.out.println(" 第 " + (start + 1) + " 页 ");
    for (Teacher teacher : teacherList) {
        System.out.println(teacher.getTch_id( ) + "," + teacher.getTch_name( ) + "," + teacher.getTch_position( ));
    }
}
```

（3）运行单元测试方法，观察控制台输出，如图 4 所示。

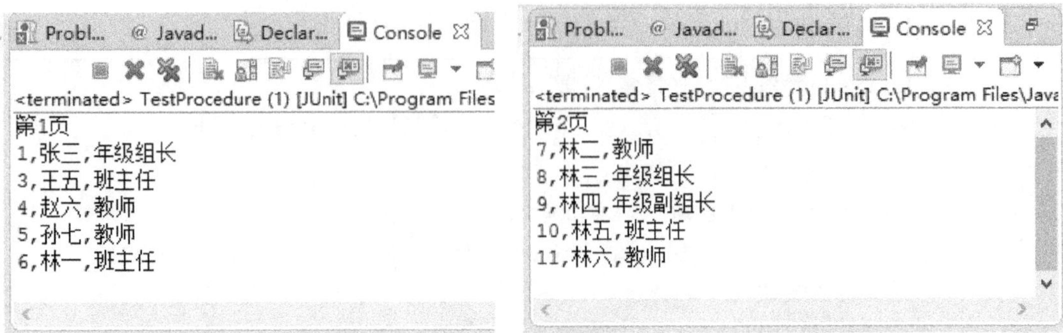

图 4　控制台输出

7.2.2　实例演示：注解配置方式

（1）增加 ITeacherMapper.java 接口，示例代码如下：

```
public interface ITeacherMapper {
@Select("select * from teacher limit #{offset},#{pagesize}")
public List<Teacher> selectTeacherInfo(Map map);
}
```

（2）配置文件，增加如下映射代码：

```
<mappers>
  <mapper resource="com/iss/mapper/teacherMapper.xml" />
  <mapper class="com.iss.mapper.ITeacherMapper" />
</mappers>
```

（3）编写测试类，示例代码如下：

```
@Test
// 注解测试方式
 public void testSqlectTeacher2( ) {
 InputStream stream = this.getClass( ).getResourceAsStream("myBatisConfig.xml");
 SqlSessionFactory factory = new SqlSessionFactoryBuilder( ).build(stream);
 SqlSession sqlSession = factory.openSession(true);
 int start = 0;// 正确的叫法应该是 offset
 int pagesize = 5;// 正确的叫法应该是 limit
 Map map = new HashMap( );
 map.put("offset", start * pagesize);
 map.put("pagesize", pagesize);
```

```
ITeacherMapper teacherMappter = sqlSession.getMapper(ITeacherMapper.class);
List<Teacher> teacherList = teacherMappter.selectTeacherInfo(map);
System.out.println(" 第 " + (start + 1) + " 页 ");
for (Teacher teacher : teacherList) {
        System.out.println(teacher.getTch_id( ) + "," + teacher.getTch_name( ) + "," + teacher.getTch_position( ));
}
}
```

（4）测试结果如图 5 所示。

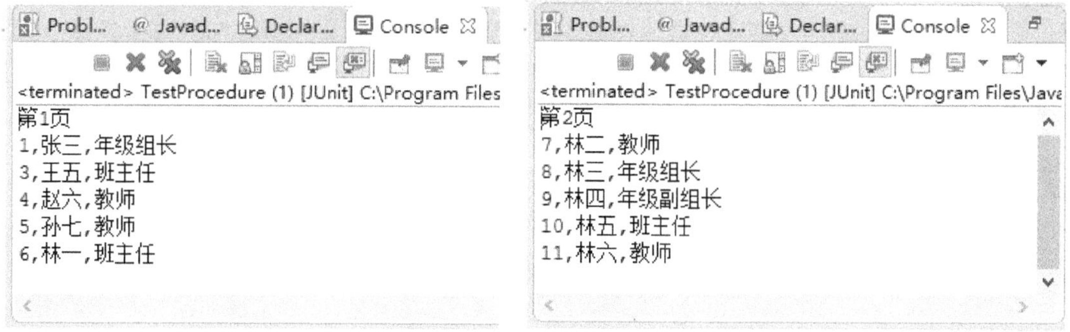

图 5　测试结果

小结

分页是最常用的功能，不管是逻辑分页还是物理分页，几乎每一个应用都离不开分页，请大家尝试各种数据库的分页方式，本书都是以 MySQL 数据库为基础的，建议大家尝试一下 Oracle 和 SQLServer 的分页方法。

经典面试题

（1）MyBatis 如何实现 MySQL 逻辑分页？

（2）MyBatis 怎样实现 MySQL 物理分页？

（3）Oracle 分页语句如何编写？

（4）SQLServer 分页语句如何编写？

（5）MyBatis 有没有自带的分页工具？

跟我上机

使用 BootStrapTable 组件结合 MyBatis 按条件查询物理分页，完成图 6 所示功能。注：数据库自行设计。

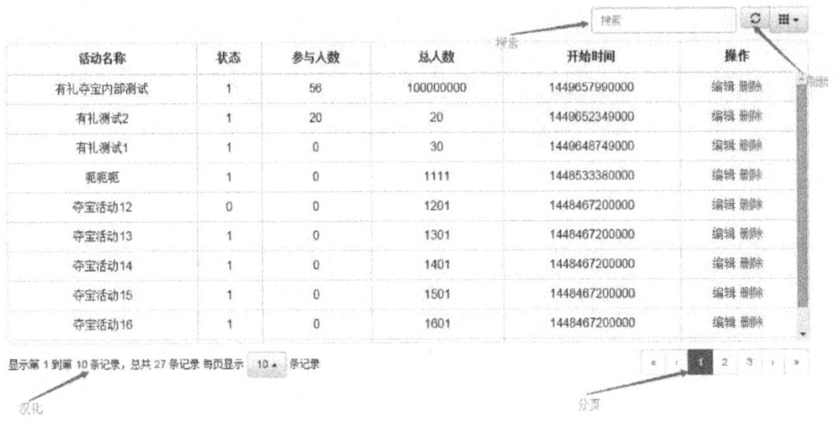

活动名称	状态	参与人数	总人数	开始时间	操作
有礼夺宝内部测试	1	56	100000000	1449657990000	编辑 删除
有礼测试2	1	20	20	1449652349000	编辑 删除
有礼测试1	1	0	30	1449648749000	编辑 删除
呃呃呃	1	0	1111	1448533380000	编辑 删除
夺宝活动12	0	0	1201	1448467200000	编辑 删除
夺宝活动13	1	0	1301	1448467200000	编辑 删除
夺宝活动14	1	0	1401	1448467200000	编辑 删除
夺宝活动15	1	0	1501	1448467200000	编辑 删除
夺宝活动16	1	0	1601	1448467200000	编辑 删除

图 6　示例功能

第 8 章　MyBatis 调用存储过程

本章要点(学会后请在方框中打钩):

- □ 学会使用 SQL 语句创建存储过程,调用存储过程
- □ 掌握使用 Mapper 配置文件方式调用存储过程
- □ 掌握使用注解配置方法调用存储过程
- □ 掌握存储过程输入参数和输出参数的用法

学会 XML 配置和注解两种配置方式的存储过程的使用，在实际开发场景中运用自如，自行尝试更为复杂的存储过程的使用（如多表查询，试题，分页等）

8.1 提出需求

查询得到男性或女性的数量，如果传入的是 0 就为女性，反之为男性。

8.2 准备数据库表和存储过程

1）创建 p_user 表

```
create table p_user(
id int primary key auto_increment,
 name varchar(10),
sex char(2)
  );
```

2）插入记录

```
insert into p_user(name,sex) values('张建军','男');
insert into p_user(name,sex) values('高雅','女');
insert into p_user(name,sex) values('李林','男');
```

3）创建存储过程

要求：查询得到男性或女性的数量，如果传入的是 0 就为女性，反之为男性。

```
DELIMITER $
CREATE PROCEDURE user_count(IN sex_id INT, OUT user_count INT)
BEGIN
IF sex_id=0 THEN
SELECT COUNT(*) FROM p_user WHERE p_user.sex='女' INTO user_count;
ELSE
SELECT COUNT(*) FROM p_user WHERE p_user.sex='男' INTO user_count;
END IF;
END
$
```

4）SQL 调用存储过程

```
DELIMITER ;
SET @user_count = 0;
```

```
CALL user_count(1, @user_count);
SELECT @user_count;
```

8.3 编辑 userMapper.xml

新建 Java 工程项目，工程目录结构如图 1 所示。

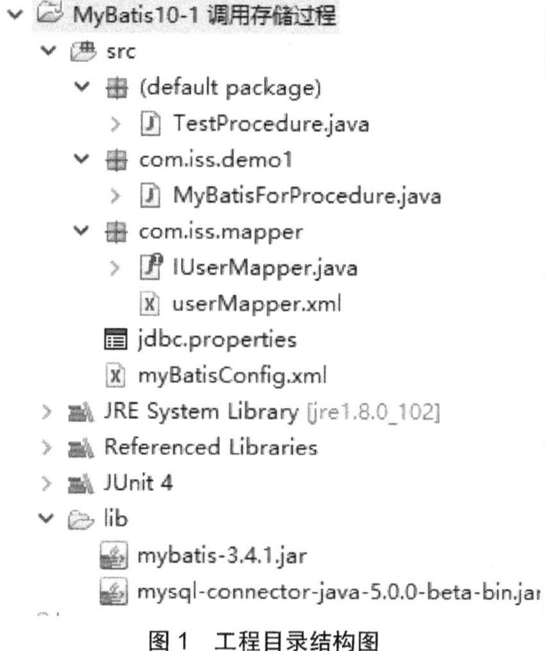

图 1　工程目录结构图

编辑 userMapper.xml 文件，添加如下配置项，具体代码内容如下。

```
<?xml version="1.0" encoding="UTF-8"?>
<!DOCTYPE mapper PUBLIC "-//mybatis.org//DTD Mapper 3.0//EN"
"http://mybatis.org/dtd/mybatis-3-mapper.dtd">
<mapper namespace="com.iss.mapper.userMapper">
<!-- 查询得到男性或女性的数量，如果传入的是 0 就女性否则是男性 -->
<select id="getUserCount" parameterMap="getUserCountMap"
        statementType="CALLABLE">
        CALL user_count(?,?)
</select>

<!— 此为对应 Java 中代码
        parameterMap.put("sexid", 0);
parameterMap.put("usercount", -1);
-->
```

```
<parameterMap type="java.util.Map" id="getUserCountMap">
        <parameter property="sexid" mode="IN" jdbcType="INTEGER" />
        <parameter property="usercount" mode="OUT" jdbcType="INTEGER" />
    </parameterMap>
</mapper>
```

8.4 编写单元测试代码

单元测试代码内容如下。

```
public class TestProcedure {
    @Test
    public void testProcedure( ) {
        InputStream stream =  this.getClass( ).getResourceAsStream("myBatisConfig.xml");
        SqlSessionFactory factory = new SqlSessionFactoryBuilder( ).build(stream);
        SqlSession sqlSession = factory.openSession(true);
        Map<String, Integer> parameterMap = new HashMap<String, Integer>( );
        parameterMap.put("sexid", 1);// 查询性别为 1 的人数
        parameterMap.put("usercount", -1);
        String statement = "com.iss.mapper.userMapper.getUserCount";
        sqlSession.selectOne(statement, parameterMap);
        Integer result = parameterMap.get("usercount");
        System.out.println(" 性别为 1 的用户人数："+result);
    }
}
```

8.5 查看测试结果

测试结果如图 2 所示。

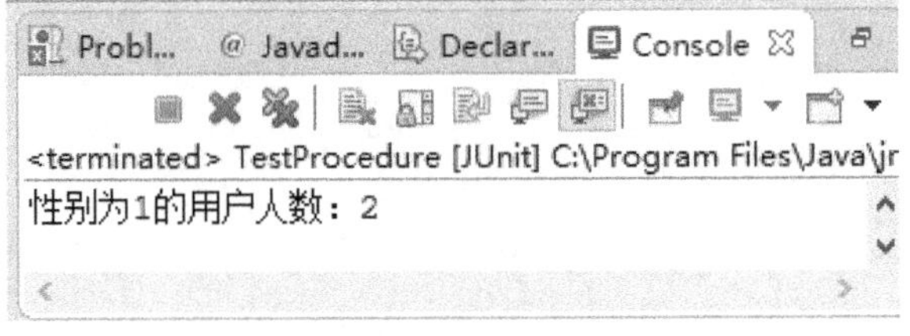

图 2 测试结果

8.6 注解配置调用存储过程

1）编写 IUserMapper.java 类

```
public interface IUserMapper {
    @Select("call    user_count(#{sexid,mode=IN,jdbcType=INTEGER},#{usercount,mode=OUT,jdbcType=INTEGER})")
    @Options(statementType = StatementType.CALLABLE)
    public Integer getUserCount(Map<String, Integer> map);
}
```

专家提示

（1）必须要写 mode 属性，同时指定哪个是输入参数，哪个是输出参数。

（2）必须要指定 jdbcType 属性。

（3）必须要写上 @Options(statementType = StatementType.CALLABLE)，指定 StatementType 是调用存储过程。

2）修改 myBatisConfig.xml

```
<mappers>
    <!-- <mapper resource="com/iss/mapper/userMapper.xml" /> -->
    <mapper class="com.iss.mapper.IUserMapper" />
</mappers>
```

3）编写单元测试代码

```
@Test
public void testProcedure2( ) {
    InputStream stream = this.getClass( ).getResourceAsStream("myBatisConfig.xml");
    SqlSessionFactory factory = new SqlSessionFactoryBuilder( ).build(stream);
    SqlSession sqlSession = factory.openSession(true);
    Map<String, Integer> parameterMap = new HashMap<String, Integer>( );
    parameterMap.put("sexid", 0);// 查询性别为 0 的人数
    parameterMap.put("usercount", -1);
    IUserMapper iusermapper = sqlSession.getMapper(IUserMapper.class);
    iusermapper.getUserCount(parameterMap);
    Integer result = parameterMap.get("usercount");
    System.out.println(" 性别为 0 的用户人数：" + result);
}
```

4）查看测试结果

测试结果如图 3 所示。

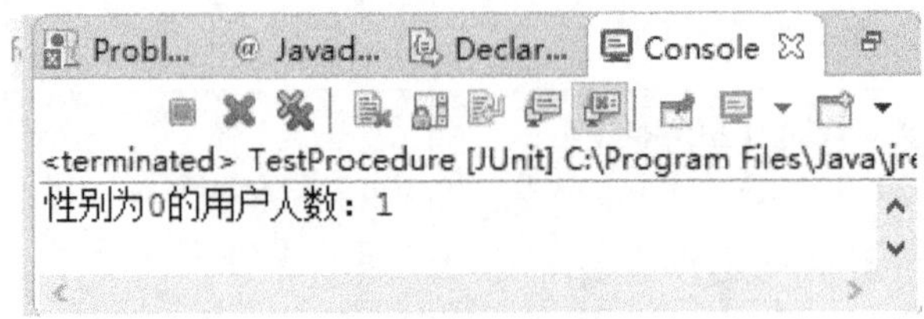

图 2 测试结果

小结

常用的操作数据库语言 SQL 语句在执行的时候需要先编译，然后执行，而存储过程（Stored Procedure）是一组为了完成特定功能的 SQL 语句集，经编译后存储在数据库中，用户通过给定名字的参数（如果该存储过程带有参数）来定义及调用它。

一个存储过程是一个可编程的函数，它在数据库中创建并保存。它可以由 SQL 语句和一些特殊的控制结构组成。当希望在不同的应用程序或平台上执行相同的函数，或者封装特定功能时，存储过程是非常有用的。数据库中的存储过程可以看作对编程中面向对象方法的模拟，它允许控制数据的访问方式。

经典面试题

（1）数据库创建存储过程的关键词是什么？

（2）MyBatis 怎么调用存储过程？

（3）MyBatis 如何设置输入参数和输出参数？

（4）MyBatis 调用存储过程如何使用数组入参？

（5）Oracle 数据库中如何编写分页存储过程？

跟我上机

（1）编写一个物理分页的存储过程 selectUserInfo，查询某个客户的信息，要求分页显示（使用输入参数为客户 ID、查看的页数、每页显示的记录数）。注：数据库表自行设计。

（2）编写一个存储过程 usp_InsertUsersData，对 userInfo 表插入用户记录。

第 9 章　MyBatis 缓存机制

本章要点（学会后请在方框中打钩）：

- □ 了解 MyBatis 缓存介绍
- □ 熟悉 MyBatis 的配置
- □ 熟练使用 MyBatis 一级缓存测试
- □ 熟练使用 MyBatis 二级缓存测试

在 MyBatis 中，缓存的功能由根接口 Cache（org.apache.ibatis.cache.Cache）定义。整个体系采用装饰器设计模式，数据存储和缓存的基本功能由 PerpetualCache（org.apache.ibatis.cache.impl.PerpetualCache）永久缓存实现，然后通过一系列的装饰器来对 PerpetualCache 永久缓存进行缓存策略等方便的控制。所以在学习过程中了解缓存机制会给我们项目带来贡献。

9.1 MyBatis 缓存介绍

正如大多数持久层框架一样，MyBatis 同样提供了一级缓存和二级缓存的支持，具体内容如下。

（1）一级缓存：基于 PerpetualCache 的 HashMap 本地缓存，其存储作用域为 Session，当 Session flush 或 close 之后，该 Session 中的所有 Cache 就将清空。

（2）二级缓存与一级缓存其机制相同，默认也是采用 PerpetualCache，HashMap 存储，不同在于其存储作用域为 Mapper(Namespace)，并且可自定义存储源，如 Ehcache。

（3）对于缓存数据更新机制，当某一个作用域 (一级缓存 Session/ 二级缓存 Namespaces) 进行了 C/U/D 操作后，默认该作用域下所有 select 中的缓存将被 clear。

9.2 MyBatis 一级缓存测试

```
/**
* 测试一级缓存
*/
public class TestOneLevelCache {
/*
* 一级缓存 : 也就 Session 级的缓存 ( 默认开启 )
*/
@Test
public void testCache1( ) {
SqlSession session = MyBatisUtil.getSqlSession( );
String statement = "com.iss.mapper.userMapper.getUser";
User user = session.selectOne(statement, 1);
System.out.println(user);
/*
* 一级缓存默认就会被使用
*/
user = session.selectOne(statement, 1);
```

```
System.out.println(user);
session.close( );
/*
1. 必须是同一个 Session, 如果 session 对象已经 close( ) 过了就不可能用了
*/
session = MyBatisUtil.getSqlSession( );
user = session.selectOne(statement, 1);
System.out.println(user);
/*
2. 查询条件是一样的
*/
user = session.selectOne(statement, 2);
System.out.println(user);
/*
3. 没有执行过 session.clearCache( ) 清理缓存
*/
//session.clearCache( );
user = session.selectOne(statement, 2);
System.out.println(user);

            /*
             4. 没有执行过增删改的操作 ( 这些操作都会清理缓存 )
             */
            session.update("com.iss.mapper.userMapper.updateUser",
                        new User(2, "user", 23));
            user = session.selectOne(statement, 2);
            System.out.println(user);
    }
}
```

9.3 MyBatis 二级缓存测试

9.3.1 开启二级缓存

在 userMapper.xml 文件中添加如下配置文件，具体代码内容如下。

```
<mapper namespace="com.iss.mapper.userMapper">
<!-- 开启二级缓存 -->
<cache/>
```

9.3.2 测试二级缓存

```
public class TestTwoLevelCache {
// 使用两个不同的 SqlSession 对象去执行相同查询条件的查询,第二次查询时不会再发送 SQL 语句,而是直接从缓存中取出数据
@Test
public void testCache2( ) {
String statement = "com.iss.mapper.userMapper.getUser";
SqlSessionFactory factory = MyBatisUtil.getSqlSessionFactory( );
// 开启两个不同的 SqlSession
SqlSession session1 = factory.openSession( );
SqlSession session2 = factory.openSession( );
// 使用二级缓存时，User 类必须实现一个 Serializable 接口 ===> User implements Serializable
User user = session1.selectOne(statement, 1);
session1.commit( );// 不懂为啥,这个地方一定要提交事务之后二级缓存才会起作用
System.out.println("user="+user);
// 由于使用的是两个不同的 SqlSession 对象,所以即使查询条件相同,一级缓存也不会开启使用
user = session2.selectOne(statement, 1);
//session2.commit( );
System.out.println("user2="+user);
}
}
```

9.3.3 二级缓存补充说明

(1)映射语句文件中的所有 select 语句将会被缓存。

(2)映射语句文件中的所有 insert,update 和 delete 语句会刷新缓存。

(3)缓存会使用 Least Recently Used(LRU,最近最少使用的)算法来收回。

(4)缓存会根据指定的时间间隔来刷新。

(5)缓存会存储 1024 个对象。

9.4 cache 标签常用属性

```
1. <cache
2. eviction="FIFO" <!-- 回收策略为先进先出 -->
3. flushInterval="60000" <!-- 自动刷新时间 60s-->
4. size="512" <!-- 最多缓存 512 个引用对象 -->
5. readOnly="true"/> <!-- 只读 -->
```

小结

MyBatis 将数据缓存设计成两级结构，分为一级缓存和二级缓存。

（1）一级缓存是 Session 会话级别的缓存，位于表示一次数据库会话的 SqlSession 对象之中，又被称之为本地缓存。一级缓存是 MyBatis 内部实现的一个特性，用户不能配置，默认情况下自动支持的缓存，用户没有定制它的权利（不过这也不是绝对的，可以通过开发插件对它进行修改）；

（2）二级缓存是 Application 应用级别的缓存，它的生命周期很长，跟 Application 的生命周期一样，也就是说它的作用范围是整个 Application 应用。

经典面试题

（1）什么是查询缓存？

（2）MyBatis 的缓存有几种？它们的工作原理都是什么？

（3）怎么使用 MyBatis 缓存机制？

（4）运用 MyBatis 进行数据库增删改时，缓存会怎样？

（5）二级缓存是什么意思？

（6）运用 Mybatis 打开二级缓存怎么配置？

第 10 章　MyBatis 日志管理

本章要点（学会后请在方框中打钩）：

- □ MyBatis 日志管理介绍
- □ LOG4J 日志工具的使用方法
- □ 打印日志到 SQL 数据库

对于以往的开发过程，我们会经常使用到 debug 模式来调节，跟踪代码执行过程。但是现在使用 MyBatis 是基于接口，配置文件的源代码执行过程。因此，我们必须选择日志工具来作为开发、调试程序的工具。

MyBatis 内置的日志工厂提供日志功能，具体的日志实现可依靠以下几种工具。

- SLF4J
- Apache Commons Logging
- Log4j 2
- Log4j
- JDK logging

具体选择哪个日志实现工具由 MyBatis 的内置日志工厂确定，它会按上文列举的顺序查找使用最先找到的工具，如果一个都未找到，日志功能将会被禁用。

通过在 MyBatis 的配置文件 mybatis-config.xml 里面添加一项 setting（配置）来选择一个不同的日志实现，具体代码内容如下。

```
<configuration>
    <settings>
        ...
        <setting name="logImpl" value="LOG4J"/>
        ...
    </settings>
</configuration>
```

logImpl 可选的值有 SLF4J、LOG4J、LOG4J2、JDK_LOGGING、COMMONS_LOGGING、STDOUT_LOGGING、NO_LOGGING 或者是实现了接口 org.apache.ibatis.logging.Log 的类的完全限定类名，并且这个类的构造函数需要以一个字符串（String 类型）为参数（可以参考 org.apache.ibatis.logging.slf4j.Slf4jImpl.java 的实现）。

```
org.apache.ibatis.logging.LogFactory.useSlf4jLogging( );
org.apache.ibatis.logging.LogFactory.useLog4JLogging( );
org.apache.ibatis.logging.LogFactory.useJdkLogging( );
org.apache.ibatis.logging.LogFactory.useCommonsLogging( );
org.apache.ibatis.logging.LogFactory.useStdOutLogging( );
```

如果的确需要调用以上的某个方法，请在其他所有 MyBatis 方法之前调用它。另外，只有在相应日志实现中存在的前提下，调用对应的方法才是有意义的，否则 MyBatis 一概忽略。如环境中并不存在 Log4J，却调用了相应的方法，MyBatis 就会忽略这一调用，代之默认的查找顺序查找日志实现。

10.1 Log4j 的使用方法

（1）创建 log4j.properties 文件，具体代码内容如下：

```
log4j.rootLogger=debug,stdout
log4j.appender.stdout=org.apache.log4j.ConsoleAppender
log4j.appender.stdout.layout=org.apache.log4j.PatternLayout
log4j.appender.stdout.layout.ConversionPattern=%5p [%t] - %m%n
```

（2）在 pom 文件中加入 log4j 依赖，具体代码内容如下：

```
<dependency>
        <groupId>log4j</groupId>
        <artifactId>log4j</artifactId>
        <version>1.2.17</version>
</dependency>
```

（3）在 main 类中加入 log 引用，具体代码内容如下：

```
package com.iss.main;
import org.apache.ibatis.session.SqlSession;
import org.apache.log4j.Logger;
// 省略一些导入的类
public class main {
        private static Logger log = Logger.getLogger(main.class);
        public static void main(String[] args) {
                SqlSession sqlSession=SqlSessionFactoryUtil.openSession( );
                UserDao userDao = sqlSession.getMapper(UserDao.class);
                String id = "admin";
                User curUser = userDao.findUserById(id);
                if(curUser!=null){
                        log.info("HelloWorld:"+curUser.getId( ));
//                      System.out.println("HelloWorld:"+curUser.getId( ));
                }
        }
}
```

（4）重新运行 main 方法，观察控制台输出。

为什么没有配置 Setting 呢？

其实如果仔细阅读上文就会发现，Mybatis 是按照顺序加载日志组件的。本例中只使用了 LOG4J，因此 MyBatis 自动使用了 LOG4J。但是如果存在多个日志，就需要配置 Setting，具体代码内容如下。

```
<?xml version="1.0" encoding="UTF-8" ?>
<!DOCTYPE configuration
PUBLIC "-//mybatis.org//DTD Config 3.0//EN"
"http://mybatis.org/dtd/mybatis-3-config.dtd">
<configuration>
    <properties resource="jdbc.properties"/>
    <settings>
        <setting name="logImpl" value="LOG4J"/>
    </settings>
    <typeAliases>
        <typeAlias alias="User" type="com.iss.entity.User"/>
    </typeAliases>
    <environments default="development">
        <environment id="development">
            <transactionManager type="JDBC" />
            <dataSource type="POOLED">
                <property name="driver" value="${jdbc.driverClassName}" />
                <property name="url" value="${jdbc.url}" />
                <property name="username" value="${jdbc.username}" />
                <property name="password" value="${jdbc.password}" />
            </dataSource>
        </environment>
    </environments>
    <mappers>
        <mapper resource="mappers/UserMapper.xml" />
    </mappers>
</configuration>
```

10.2 综合案例演示

10.2.1 在 pom.xml 中添加依赖

```
<dependency>
    <groupId>org.slf4j</groupId>
    <artifactId>slf4j-log4j12</artifactId>
    <version>1.7.2</version>
</dependency>
```

下载如图 1 所示的 3 个 jar 包。

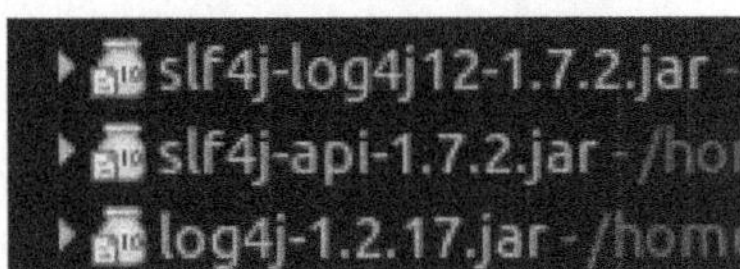

图 1 jar 包

如果不使用 maven 管理 jar 包，则下载这三个 jar 包并加载到项目 classpath 路径下即可。

10.2.2 建立 log4j.properties 文件

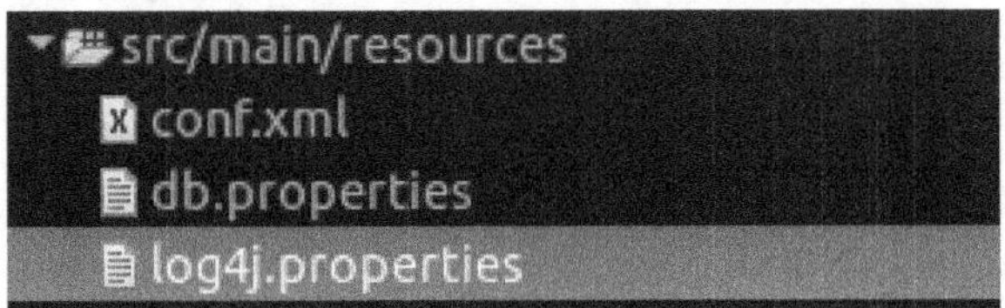

图 2 log4j.properties 文件

配置 Log4J 比较简单，比如需要记录这个 mapper 接口的日志，具体代码内容如下。

```
# Global logging configuration
log4j.rootLogger=ERROR, stdout
# MyBatis logging configuration...
log4j.logger.org.mybatis.example.BlogMapper=TRACE
# Console output...
log4j.appender.stdout=org.apache.log4j.ConsoleAppender
log4j.appender.stdout.layout=org.apache.log4j.PatternLayout
log4j.appender.stdout.layout.ConversionPattern=%5p [%t] - %m%n
```

10.2.3 配置 mapper 接口

配置 Log4J 比较简单，比如需要记录这个 mapper 接口的日志，具体代码内容如下：

```
package org.mybatis.example;
public interface BlogMapper {
    @Select("SELECT * FROM blog WHERE id = #{id}")
    Blog selectBlog(int id);
}
```

添加以上配置后，Log4J 就会把 org.mybatis.example.BlogMapper 的详细执行日志记录下来，对于应用中的其他类则仅仅记录错误信息。也可以将日志从整个 mapper 接口级别调整到语句级别，从而实现更细粒度的控制。如下配置只记录 selectBlog 语句的日志。

```
log4j.logger.org.mybatis.example.BlogMapper.selectBlog=TRACE
```

与此相对，可以对一组 mapper 接口记录日志，只要对 mapper 接口所在的包开启日志功能即可。

```
log4j.logger.org.mybatis.example=TRACE
```

某些查询可能会返回大量数据，只想记录其执行的 SQL 语句该怎么办？为此，MyBatis 中 SQL 语句的日志级别被设置为 DEBUG（JDK Logging 中为 FINE），结果日志的级别为 TRACE（JDK Logging 中为 FINER）。所以，只要将日志级别调整为 DEBUG 即可达到目的。

```
log4j.logger.org.mybatis.example=DEBUG
```

要记录日志的是类似下面的 mapper 文件而不是 mapper 接口又该怎么呢？

```
<?xml version="1.0" encoding="UTF-8" ?>
<!DOCTYPE mapper
  PUBLIC "-//mybatis.org//DTD Mapper 3.0//EN"
  "http://mybatis.org/dtd/mybatis-3-mapper.dtd">
<mapper namespace="org.mybatis.example.BlogMapper">
  <select id="selectBlog" resultType="Blog">
    select * from Blog where id = #{id}
  </select>
</mapper>
```

对这个文件记录日志，只要对命名空间增加日志记录功能即可：

```
log4j.logger.org.mybatis.example.BlogMapper=TRACE
```

要记录具体语句的日志可以这样做：

```
log4j.logger.org.mybatis.example.BlogMapper.selectBlog=TRACE
```

小结

MyBatis 通过日志工厂提供日志信息，MyBatis 内置的日志模版是 LOG4J，commons.log，jdk log 也可以通过 slf4j 简单日志模版结合 LOG4J 使用日志信息输出，具体选择哪个日志实现由 MyBatis 的内置日志工厂确定。它会使用最先找到的（按上文列举的顺序查找）。如果一个都未找到，日志功能就会被禁用。不少应用服务器的 classpath 中已经包含 Commons Logging，如 Tomcat，所以 MyBatis 会把它作为具体的日志实现。

经典面试题

（1）如何控制 MyBatis 的 SQL 日志输出？
（2）如何关闭 MyBatis 日志？
（3）如何使用 MyBatis 与 Spring 整合使用 LOG4J 打印日志到控制台？
（4）常用的日志工具有哪些？
（5）使用 Spring MVC+MyBatis 实现日志处理记录到数据库？

跟我上机

配置 log4j.properties 配置文件，在控制台打印出 MyBatis 运行的 SQL 语句？
配置文件提示如下。

```
1. 设置 Logger 输出级别和输出目的地 ###
2. log4j.rootLogger=debug,stdout,logfile
3. ### 把日志信息输出到控制台 ###
4. log4j.appender.stdout=org.apache.log4j.ConsoleAppender
5. #log4j.appender.stdout.Target=System.err
6. log4j.appender.stdout.layout=org.apache.log4j.SimpleLayout
7. ### 把日志信息输出到文件：iss.log ###
8. log4j.appender.logfile=org.apache.log4j.FileAppender
9. log4j.appender.logfile.File=iss.log
10. log4j.appender.logfile.layout=org.apache.log4j.PatternLayout
11. log4j.appender.logfile.layout.ConversionPattern=%d{yyyy-MM-dd HH:mm:ss} %F %p %m%n
12. ### 显示 SQL 语句部分 ###
```

```
log4j.logger.com.ibatis=DEBUG
log4j.logger.com.ibatis.common.jdbc.SimpleDataSource=DEBUG
log4j.logger.com.ibatis.common.jdbc.ScriptRunner=DEBUG
log4j.logger.com.ibatis.sqlmap.engine.impl.SqlMapClientDelegate=DEBUG
log4j.logger.java.sql.Connection=DEBUG
log4j.logger.java.sql.Statement=DEBUG
log4j.logger.java.sql.PreparedStatement=DEBUG
```

附录　Spring+Spring MVC+MyBatis 全注解整合

本案例实现了三国英雄在线投票的功能，功能包括投票功能、删除功能、添加功能、上传头像功能和查看英雄人物简介功能。注：UI 采用 BootStrap 框架实现。

1. 在线投票系统实现功能截图

在线投票系统和查看英雄人物简介如图 1 所示，上传头像和添加功能如图 2 所示。

三国英雄投票

投票	编号	头像	姓名	票数	百分比
	1		孙权	54	25.7%
	2		刘表	25	11
	3		马谡	31	14
	4		郭嘉	100	47.6%

投票　删除　添加英雄　　　方框表示可单击区域

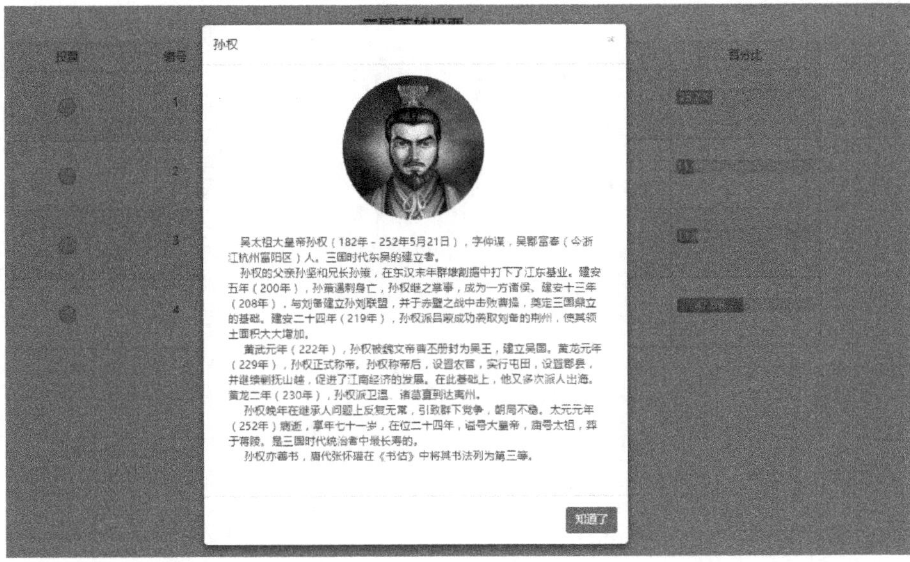

图 1　细节图

图 2　细节图

2. 项目工程结构截图

项目工程结构截图如图 3 所示。

- SSMTouPiao
  - JAX-WS Web Services
  - Java Resources
    - src
      - com.iss
        - UUIDUtils.java
      - com.iss.controller
        - MyController.java
      - com.iss.mapper
        - TouPiaoMapper.java
      - com.iss.pojo
        - Characters.java
      - com.iss.service
        - TouPiaoService.java
      - com.iss.service.impl
        - TouPiaoServiceImpl.java
      - jdbcConfig.properties
  - JavaScript Resources
  - build
  - WebContent
    - bootstrap
    - css
    - fileinput
    - fileUpload
    - img
    - js
    - META-INF
    - wangediter
    - WEB-INF
      - lib
      - applicationContext.xml
      - spring-mvc.xml
      - web.xml
    - index.jsp

图 3　项目工程结构图

3. 所需要的包截图

所需要的包的截图如图 4 所示。

图 4 所需要的包的截图

4.web.xml 部署描述器文件配置

web.xml 部署描述器文件配置具体代码如下所示。

```
<?xml version="1.0" encoding="UTF-8"?>
<web-app
    xmlns="http://xmlns.jcp.org/xml/ns/javaee"  xmlns:xsi="http://www.w3.org/2001/XMLSchema-instance"
        xsi:schemaLocation="http://xmlns.jcp.org/xml/ns/javaee
    http://xmlns.jcp.org/xml/ns/javaee/web-app_3_1.xsd"version="3.1">
        <context-param>
            <param-name>contextConfigLocation</param-name>
            <param-value>WEB-INF/applicationContext.xml</param-value>
        </context-param>
        <listener>
        <listener-class>org.springframework.web.context.ContextLoaderListener</listener-class>
        </listener>
```

```
    <servlet>
        <servlet-name>springmvc</servlet-name>
        <servlet-class>org.springframework.web.servlet.DispatcherServlet</servlet-class>
        <init-param>
            <param-name>contextConfigLocation</param-name>
            <param-value>WEB-INF/spring-mvc.xml</param-value>
        </init-param>
        <load-on-startup>1</load-on-startup>
        <async-supported>true</async-supported>
    </servlet>
    <servlet-mapping>
        <servlet-name>springmvc</servlet-name>
        <url-pattern>/</url-pattern>
    </servlet-mapping>
</web-app>
```

5.jdbcConfig.properties 数据库配置文件

```
driverClassName=org.gjt.mm.mysql.Driver
url=jdbc:mysql://localhost:3306/studentdb?useUnicode=true&characterEncoding=utf-8
username=root
password=root
```

6.applicationContext.xml Spring 核心配置文件

```
<?xml version="1.0" encoding="UTF-8"?>
<beans xmlns="http://www.springframework.org/schema/beans"
xmlns:xsi="http://www.w3.org/2001/XMLSchema-instance"
xsi:schemaLocation="http://www.springframework.org/schema/beans http://www.springframework.org/schema/beans/spring-beans.xsd">
<bean id="jdbc" class="org.springframework.beans.factory.config.PropertyPlaceholderConfigurer">
    <property name="locations" value="classpath*:jdbcConfig.properties" />
</bean>
<bean id="dataSource"
    class="org.springframework.jdbc.datasource.DriverManagerDataSource">
    <property name="driverClassName" value="${driverClassName}" />
    <property name="url" value="${url}" />
```

```
        <property name="username" value="${username}" />
        <property name="password" value="${password}" />
</bean>
<!-- 开启事务 -->
<bean id="trans"
        class="org.springframework.jdbc.datasource.DataSourceTransactionManager">
        <property name="dataSource" ref="dataSource" />
</bean>
<bean id="sqlsession" class="org.mybatis.spring.SqlSessionFactoryBean">
        <property name="dataSource" ref="dataSource" />
</bean>
<bean id="mapperScanner" class="org.mybatis.spring.mapper.MapperScannerConfigurer">
        <property name="basePackage" value="com.iss.mapper" />
</bean>
</beans>
```

7.spring-mvc.xml Spring MVC 核心配置文件

```
<?xml version="1.0" encoding="UTF-8"?>
<beans xmlns="http://www.springframework.org/schema/beans"
xmlns:xsi="http://www.w3.org/2001/XMLSchema-instance"    xmlns:context="http://www.springframework.org/schema/context"
xmlns:mvc="http://www.springframework.org/schema/mvc"
xsi:schemaLocation="http://www.springframework.org/schema/beans
  http://www.springframework.org/schema/beans/spring-beans.xsd
  http://www.springframework.org/schema/context
  http://www.springframework.org/schema/context/spring-context.xsd
  http://www.springframework.org/schema/mvc
  http://www.springframework.org/schema/mvc/spring-mvc.xsd">
<context:component-scan base-package="com.iss" />
<mvc:annotation-driven>
        <mvc:message-converters>
                <bean
    class="com.alibaba.fastjson.support.spring.FastJsonHttpMessageConverter" />
        </mvc:message-converters>
</mvc:annotation-driven>
<bean id="multipartResolver"
        class="org.springframework.web.multipart.commons.CommonsMultipartResolver">
```

```
        <property name="maxUploadSize">
            <value>10000000</value>
        </property>
        <property name="defaultEncoding">
            <value>utf-8</value>
        </property>
</bean>
<mvc:default-servlet-handler />
</beans>
```

8.MyController.java 控制器

```
package com.iss.controller;
// 省略导入类
@Controller
@Scope("prototype")
public class MyController {
        @Autowired
        TouPiaoService touPiaoServiceImpl;

        @RequestMapping("findAll")
        @ResponseBody
        public List<Characters> findAllChaInfo( ){
            int sum=0;
            List<Characters> characterList=touPiaoServiceImpl.findAllChaInfo( );
            for (Characters c:characterList) {
                sum+=c.getNumber( );
            }
            for (Characters c:characterList) {
                c.setSum(sum);
            }
            return characterList;
        }

        @RequestMapping("submit")
        @ResponseBody
        public int TouPiao(int id){
            int i= touPiaoServiceImpl.TouPiao(id);
```

```
            System.out.println(i);
            return i;
        }

    @RequestMapping("ImgUpload")
    @ResponseBody
    public String ImgUpload(HttpServletRequest request,int id,@RequestParam MultipartFile myImg){
            String realPath = request.getSession( ).getServletContext( ).getRealPath("/fileUpload");
            String uploadFilePath = realPath+"/"+myImg.getOriginalFilename( );
            File file = new File(uploadFilePath);
            int i=0;
            try {
                myImg.transferTo(file);
                i = touPiaoServiceImpl.ImgUpload(id, "fileUpload/" + myImg.getOriginalFilename( ));
                if(i>0) {
                    System.out.println(" 数据库记录成功 ");
                }else {
                    System.out.println(" 数据库记录失败 ");
                }
               return "true";
            } catch (Exception e) {
                e.printStackTrace( );
            }
            return "asdasd";
    }

    @RequestMapping("findById")
    @ResponseBody
    public Characters findChaInfoById(int id) {
        Characters c = touPiaoServiceImpl.findChaInfoById(id);
        return c;
    }
```

```
    @RequestMapping("deleteChaById")
    @ResponseBody
    public int deleteChaById(int id) {
        int i = touPiaoServiceImpl.deleteChaById(id);
        return i;
    }

    @RequestMapping("addChaInfo")
    @ResponseBody
    public int addChaInfo(int id, String sname, String resume) {
        int i=touPiaoServiceImpl.addChaInfo(id, sname, resume);
        return i;
    }
}
```

9.TouPiaoService.java 服务层接口

```
package com.iss.service;
import com.iss.pojo.Characters;
import java.util.List;
public interface TouPiaoService {
    public List<Characters> findAllChaInfo( );
    public int TouPiao(int id);
    public int ImgUpload(int id, String myimg);
    public Characters findChaInfoById(int id);
    public int deleteChaById(int id);
    public int addChaInfo(int id, String sname, String resume);
}
```

10.TouPiaoServiceImpl.java

```
package com.iss.service.impl;
// 省略导入类
@Service
public class TouPiaoServiceImpl implements TouPiaoService {
@Autowired
TouPiaoMapper touPiaoMapper;
@Override
```

```
public List<Characters> findAllChaInfo( ) {
        List<Characters> characterList = touPiaoMapper.findAllChaInfo( );
        return characterList;
}

@Override
public int TouPiao(int id) {
        int i = touPiaoMapper.TouPiao(id);
        return i;
}

@Override
public int ImgUpload(int id, String myimg) {
        int i = touPiaoMapper.ImgUpload(id, myimg);
        return i;
}

@Override
public Characters findChaInfoById(int id) {
        Characters c = touPiaoMapper.findChaInfoById(id);
        return c;
}

@Override
public int deleteChaById(int id) {
        int i = touPiaoMapper.deleteChaById(id);
        return i;
}

@Override
public int addChaInfo(int id, String sname, String resume) {
        int i = touPiaoMapper.addChaInfo(id, sname, resume);
        return i;
}
}
```

11.TouPiaoMapper.java Mapper 接口

```
package com.iss.mapper;
// 省略导入类
@Repository
public interface TouPiaoMapper {
@Select("SELECT * from tb_toupiao")
public List<Characters> findAllChaInfo( );

@Update("UPDATE tb_toupiao set number=number+1 where id=#{id}")
public int TouPiao(int id);

@Update("update tb_toupiao set myimg=#{myimg} where id=#{id}")
public int ImgUpload(@Param("id") int id, @Param("myimg") String myimg);

@Select("select * from tb_toupiao where id=#{id}")
public Characters findChaInfoById(int id);

@Delete("delete from tb_toupiao where id=#{id}")
public int deleteChaById(int id);

@Insert("insert into tb_toupiao (id,sname,resume) values (#{id},#{sname},#{resume})")
public int addChaInfo(@Param("id") int id, @Param("sname") String sname, @Param("resume") String resume);
}
```

12.Characters.java 实体类

```
package com.iss.pojo;
public class Characters {
int id;
String sname;
int number;
String resume;
String myimg;
int sum;
// 省略 setter 和 getter
}
```

13.index.jsp 投票界面

```
<%@ page contentType="text/html;charset=UTF-8" language="java" %>
<html>
<head>
    <link rel="stylesheet" href="bootstrap/css/bootstrap.min.css">
    <link rel="stylesheet" href="fileinput/fileinput.min.css">
    <link rel="stylesheet" href="css/style.css">
    <script type="text/javascript" src="js/jquery-3.2.1.min.js"></script>
    <script type="text/javascript" src="bootstrap/js/bootstrap.min.js"></script>
    <script type="text/javascript" src="js/jquery.form.js"></script>
    <script type="text/javascript" src="fileinput/fileinput.min.js"></script>
    <script type="text/javascript" src="fileinput/fileinput_locale_zh.js"></script>
    <script type="text/javascript" src="wangediter/wangEditor.js"></script>
    <title> 投票 </title>
    <script>
        var sid = 0;
        var maxRowNum;
        $(document).ready(function ( ) {
            findAllChaInfo( );
        });
        function findAllChaInfo( ) {
            $.ajax({
                url: "findAll",
                method: "post",
                success: function (result) {
                    var str = "";
                    var errorUrl = "img/error.png";
                    $.each(result, function (index, obj) {
                        str += "<tr><td>" + "<input class='myRadio'  type='radio' name='radio' onchange='javascrit:radioTest( )' value=" + obj.id + ">"
                            + "</td><td ><p>" + obj.id + "</p></td><td>" + "<img name='showImg' src='" + obj.myimg + "' onclick='javascript:showModal(" + obj.id + ")' onerror='this.src=\"img/error.png\"' class='img-circle showImg' width='80px' height='80px'>" + "</td><td class='myId'>" +
                            "<p id='sname'  onclick='showHeroResumeModel(" + obj.id + ")'>" + obj.sname + "</p></td><td>"
```

```
                                        + obj.number + "</td><td>"
                                        + "<div class='progress'><div class='progress-bar progress-bar-success progress-bar-striped active' role='progressbar' aria-valuenow=" + obj.number +
                                        " aria-valuemin='0' aria-valuemax=" + obj.sum + " style='width: " + (obj.number / obj.sum) * 100 + "%'> " +
                                        "<span class='sr-only'> 您的浏览器不支持 JS?</span>" +
                                        +((obj.number / obj.sum) * 100).toFixed(1) + "%</div>" + "</td></tr>"
                                maxRowNum = index + 1;
                            });
                            $("#tbody").html(str);
                        }
                    })
                }

                /* 点击 radio 获得本行 id*/
                function radioTest( ) {
                    sid = $('input:radio:checked').val( );
                }
                /* 提交投票 */
                function submit( ) {
                    $.ajax({
                        url: "submit",
                        method: "post",
                        data: {
                            "id": sid
                        },
                        success: function ( ) {
                            $('#myModal3').modal('show');
                            findAllChaInfo( );
                        },
                        error: function ( ) {
                            alert(" 请选择一行 ");
                        }
                    })
```

```
        }
        /* 删除英雄 */
        function deleteChaById( ) {
            if (sid == 0) {
                alert(" 请选择您要删除的对象！ ");
            } else {
                var v = confirm(" 确定要删除 ?");
                if (v == true) {
                    $.ajax({
                        url: "deleteChaById",
                        method: "post",
                        data: {
                            "id": sid
                        },
                        success: function (result) {
                            if (result > 0) {
                                findAllChaInfo( );
                                alert(" 删除成功 ~")
                            } else
                                alert(" 删除失败！ ")
                        },
                        error: function ( ) {
                            alert(" 出错！ ");
                        }
                    });
                }
            }
        }
        /* 显示上传图片模块框 */
        function showModal(id) {
            $('#myModal').modal('show');
            ImgUpload(id);
        }
        /* 显示英雄详情模块框 */
        function showHeroResumeModel(id) {
```

```
        $.ajax({
            url: "findById",
            method: "post",
            data: {
                "id": id
            },
            success: function (result) {
                $('#myModalLabel2').html(result.sname);
                $('#modal-img').attr("src", result.myimg);
                $('#personResume').html(result.resume);
            }
        });
        $('#myModal2').modal('show');
    }
    /* 文件上传事件 */
    function ImgUpload(id) {
        $('.fileinput-upload-button').click(function ( ) {
            $('#uploadImg').ajaxSubmit({
                url: "ImgUpload",
                type: "post",
                dataType: "json",
                async: false,
                data: {
                    "id": id
                },
        contentType: "application/x-www-form-urlencoded; charset=utf-8",
                success: function (result) {
                    alert(result);
                },
                error: function (data, status, e) {
                    /*  var obj=result;*/
                    alert(status);
                }
            });
        });
```

```
        }

        /* 添加新英雄事件 */
        function addChaInfo( ) {
            var sname = $('#add_Name').val( );
            var resume = editor.txt.text( );
            $.ajax({
                url: "addChaInfo",
                method: "post",
                data: {
                    "id": maxRowNum + 1,
                    "sname": sname,
                    "resume": resume
                },
                success: function (result) {
                    if (result == 1) {
                        $('#myModal4').modal('hide');
                        findAllChaInfo( );
                    } else
                        alert(" 添加失败 ");
                },
                error: function ( ) {
                    alert(" 出错！ ");
                }
            });
        }

    </script>
</head>
<body>
<%--container--%>
<div class="container">
    <div class="row">
        <div class="col-sm-12">
            <div class="center-block" style="text-align: center">
```

```
                    <h3><strong> 三国英雄投票 </strong></h3>
                <form>
                    <table align="center" class="table table-striped table-bordered">
                        <tr>
                            <td> 投票 </td>
                            <td> 编号 </td>
                            <td> 头像 </td>
                            <td> 姓名 </td>
                            <td> 票数 </td>
                            <td> 百分比 </td>
                        </tr>
                        <tbody id="tbody"></tbody>
                    </table>
                </form>
            </div>
        </div>
        <div class="col-sm-12">
            <div class="center-block">
    <center><a id="btn-submit" href="javascript:submit( )" class="btn btn-primary"> 投票 </a>
        <a href="javascript:deleteChaById( )" class="btn btn-danger"> 删除 </a>
        <a data-toggle="modal" data-target="#myModal4" class="btn btn-warning"> 添加英雄 </a></center>
            </div>
        </div>
    </div>
</div>
<%--/container--%>

<!-- Modal -->
<div class="modal fade" id="myModal" tabindex="-1" role="dialog" aria-labelledby="myModalLabel">
    <div class="modal-dialog" role="document">
        <div class="modal-content">
            <div class="modal-header">
```

```
                    <button type="button" class="close" data-dismiss="modal" aria-label="Close"><span aria-hidden="true">&times;</span>
                    </button>
                    <h4 class="modal-title" id="myModalLabel"> 上传头像 </h4>
                </div>
                <div class="modal-body">
            <form id="uploadImg" name="uploadImg" enctype="multipart/form-data">
                <input type="file" id="myImg" name="myImg" class="file" multiple><br>
                    </form>
                </div>
                <div class="modal-footer">
            <center>@Copyright © 2001- 2017 iSoftStone. All Rights Reserved</center>
                </div>
            </div>
        </div>
</div>

<%-- 英雄简介模式窗口 --%>
<div class="modal fade" id="myModal2" tabindex="-1" role="dialog" aria-labelledby="myModalLabel2">
        <div class="modal-dialog" role="document">
            <div class="modal-content">
                <div class="modal-header">
                    <button type="button" class="close" data-dismiss="modal" aria-label="Close"><span aria-hidden="true">&times;</span>
                    </button>
                    <h4 class="modal-title" id="myModalLabel2"></h4>
                </div>
                <div class="modal-body">
                    <img id="modal-img" src="" class="img-circle center-block" onerror="this.src='img/error.png'"  width="200px" height="200px">
                    <p id="personResume"></p>
                </div>
                <div class="modal-footer">
```

```
                        <button type="button" class="btn btn-primary" data-dismiss="modal"> 知道了 </button>
                    </div>
                </div>
            </div>
</div>
<%--/HeroResume--%>

<!-- 提交成功提示模块 -->
<div class="modal fade" id="myModal3" tabindex="-1" role="dialog" aria-labelledby="myModalLabel3">
    <div class="modal-dialog" role="document">
        <div class="modal-content">
            <div class="modal-header">
                <button type="button" class="close" data-dismiss="modal" aria-label="Close"><span aria-hidden="true">&times;</span>
                </button>
                <h4 class="modal-title" id="myModalLabel3"> 消息提示 </h4>
            </div>
            <div class="modal-body">
                恭喜您！投票成功！
            </div>
            <div class="modal-footer">
                <button type="button" class="btn btn-primary" data-dismiss="modal"> 关闭 </button>
            </div>
        </div>
    </div>
</div>
<!--/ 提交成功提示模块 -->

<!-- Modal -->
<div class="modal fade bs-example-modal-lg" id="myModal4" tabindex="-1" role="dialog" aria-labelledby="myModalLabel4">
    <div class="modal-dialog  modal-lg" role="document">
```

```
            <div class="modal-content">
                <div class="modal-header">
                    <button type="button" class="close" data-dismiss="modal" aria-label="Close"><span aria-hidden="true">&times;</span>
                    </button>
                    <h3 class="modal-title" id="myModalLabel4"><strong> 添加新角色 </strong></h3>
                </div>
                <div class="modal-body">
                    <form id="add_Form">
                        <input type="text" id="add_Name" required placeholder=" 请输入新角色姓名 ">
                        <div id="editor">
                            <p> 请为 <b> 角色 </b> 添加简介 </p>
                        </div>
                    </form>
                </div>
                <div class="modal-footer">
    <button type="button" class="btn btn-default" data-dismiss="modal"> 关闭 </button>
    <button type="button" class="btn btn-primary" onclick="javscript:addChaInfo( )"> 保存 </button>
                </div>
            </div>
        </div>
</div>
</body>
<!-- wangediter-->
<script type="text/javascript">
    var E = window.wangEditor;
    var editor = new E('#editor')
    // 或者 var editor = new E( document.getElementById('#editor') )
    editor.create( )
</script>
</html>
```

14.tb_toupiao 数据库表结构

tb_toupiao 表结构如图 5 所示。

名	类型	长度	小数点	不是 null	
id	int	11	0	☑	1
sname	varchar	255	0	☑	
number	int	11	0	☐	
resume	varchar	1000	0	☐	
myimg	varchar	255	0	☐	

图 5　tb_toupiao 表结构

至此，在线投票功能代码粘贴完成，以上代码仅供参考。